BIOETHICS CRITICALLY RECONSIDERED

Philosophy and Medicine

VOLUME 100

Founding Co-Editor
Stuart F. Spicker

Senior Editor

H. Tristram Engelhardt, Jr., *Department of Philosophy, Rice University, and Baylor College of Medicine, Houston, Texas*

Associate Editor

Lisa M. Rasmussen, *Department of Philosophy, University of North Carolina at Charlotte, Charlotte, North Carolina*

Assistant Editor

Jeffrey P. Bishop, *Gnaegi Center for Health Care Ethics, Saint Louis University, St. Louis, Missouri*

Editorial Board

George J. Agich, *Department of Philosophy, Bowling Green State University, Bowling Green, Ohio*
Nicholas Capaldi, *College of Business Administration, Loyola University, New Orleans, Louisiana*
Edmund Erde, *University of Medicine and Dentistry of New Jersey, Stratford, New Jersey*
Christopher Tollefsen, *Department of Philosophy, University of South Carolina, Columbia, South Carolina*
Kevin Wm. Wildes, S.J., *President Loyola University, New Orleans, Louisiana*

For further volumes:
http://www.springer.com/series/6414

BIOETHICS CRITICALLY RECONSIDERED

Having Second Thoughts

Edited by
H. TRISTRAM ENGELHARDT, JR.
Rice University, Houston, TX, USA

Editor
H. Tristram Engelhardt, Jr. Ph.D.
Main St 6100
77005-1827 HOUSTON Texas
USA
htengelh@rice.edu

ISSN 0376-7418
ISBN 978-94-007-2243-9   e-ISBN 978-94-007-2244-6
DOI 10.1007/978-94-007-2244-6
Springer Dordrecht Heidelberg London New York

Library of Congress Control Number: 2011938240

© Springer Science+Business Media B.V. 2012
No part of this work may be reproduced, stored in a retrieval system, or transmitted in any form or by
any means, electronic, mechanical, photocopying, microfilming, recording or otherwise, without written
permission from the Publisher, with the exception of any material supplied specifically for the purpose
of being entered and executed on a computer system, for exclusive use by the purchaser of the work.

Printed on acid-free paper

Springer is part of Springer Science+Business Media (www.springer.com)

Contents

1 A Skeptical Reassessment of Bioethics 1
H. Tristram Engelhardt, Jr.

Part I   History of Bioethics: Four Perspectives

2 Beginning Bioethics 31
Michael S. Yesley

3 Genesis of a Totalizing Ideology: Bioethics' Inner Hippie 49
Griffin Trotter

**4 Bioethics and Professional Medical Ethics: Mapping
and Managing an Uneasy Relationship** 71
Laurence B. McCullough

**5 Two Rival Understandings of Autonomy, Paternalism,
and Bioethical Principlism** 85
Aaron E. Hinkley

**Part II   The Practice of Bioethics and Clinical Ethics
Consultation: Three Views**

6 Bioethics as Political Ideology 99
Mark J. Cherry

7 The "s" in Bioethics: Past, Present and Future 123
Ana S. Iltis and Adrienne Carpenter

**8 Why Clinical Bioethics So Rarely Gives Morally Normative
Guidance** 151
H. Tristram Engelhardt, Jr.

Part III   The Incredible Search for Bioethical Professionalism: Some Final Critical Reflections on Circular Thinking

9   On the Social Construction of Health Care Ethics Consultation . .   177
Jeffrey P. Bishop

Index 191

Contributors

Jeffrey P. Bishop Albert Gnaegi Center for Health Care Ethics, Saint Louis University, Salus Center, Room 527, 3545 Lafayette Ave, St. Louis, MO 63104, USA, jbisho12@slu.edu

Adrienne Carpenter Albert Gnaegi Center for Health Care Ethics, Saint Louis University, St. Louis, MO 63104, USA, acarpen6@slu.edu

Mark J. Cherry The Dr. Patricia A. Hayes Professor in Applied Ethics, Professor of Philosophy, Department of Philosophy, St. Edward's University, Austin, TX 78704, USA, markc@stedwards.edu

H. Tristram Engelhardt, Jr. Department of Philosophy, MS-14, Rice University, Houston, TX 77005, USA, htengelh@rice.edu

Aaron E. Hinkley Department of Philosophy, MS-14, Rice University, Houston, TX 77005, USA, hinkley@rice.edu

Ana S. Iltis Center for Bioethics, Health and Society, Wake Forest University, Winston-Salem, NC 27109, USA, iltisas@wfu.edu

Laurence B. McCullough Dalton Tomlin Chair in Medical Ethics and Health Policy, Baylor College of Medicine, Center for Medical Ethics and Health Policy, Houston, TX 77030-2411, USA, mccullou@bcm.edu

Griffin Trotter Albert Gnaegi Center for Health Care Ethics, Salus Center, Saint Louis University, St. Louis, MO 63104, USA, trotterc@slu.edu

Michael S. Yesley Independent Consultant, Bainbridge Island, WA 98110, USA, myesley@earthlink.net

Notes on Contributors

Jeffrey P. Bishop, M.D., Ph.D., is director of the Albert Gnaegi Center for Health Care Ethics and Tenet Chair of Health Care Ethics at Saint Louis University, St. Louis, Missouri.

Adrienne Carpenter, is professor at the Albert Gnaegi Center for Health Care Ethics at Saint Louis University, St. Louis, Missouri.

Mark J. Cherry, Ph.D., is the Dr. Patricia A. Hayes Professor in Applied Ethics, Department of Philosophy, St. Edward's University Austin, Texas.

H. Tristram Engelhardt, Jr., Ph.D., M.D., is professor, department of philosophy, Rice University, and professor emeritus, departments of medicine and community medicine, Baylor College of Medicine, Houston, Texas.

Aaron E. Hinkley, ABD, is a doctoral candidate in the department of philosophy, Rice University, Houston, Texas and an adjunct professor of philosophy at Houston Baptist University. He is also the senior managing editor of *The Journal of Medicine and Philosophy.*

Ana S. Iltis, Ph.D., is Director of the Center for Bioethics, Health and Society and associate professor of philosophy at Wake Forest University in Winston-Salem, North Carolina.

Laurence B. McCullough, Ph.D., is the associate director for Education and inaugural holder of the Dalton Tomlin Chair in Medical Ethics and Health Policy in the Center for Medical Ethics and Health Policy, Baylor College of Medicine, where is he also professor of Medicine and Medical Ethics.

Griffin Trotter, M.D., Ph.D., is professor at the Center for Health Care Ethics at Saint Louis University, St. Louis, Missouri. He also has an appointment in the Department of Surgery, Emergency Medical Division in School of Medicine at Saint Louis University.

Michael S. Yesley, J.D., is an attorney and writes about bioethics, particularly issues in genetics and research with human subjects. He recently retired from Los Alamos National Laboratory (LANL).

Chapter 1
A Skeptical Reassessment of Bioethics

H. Tristram Engelhardt, Jr.

1.1 What Is Bioethics, After All: Claims for Moral Expertise in the Face of Intractable Moral Pluralism

What is bioethics? Who is a bioethicist? Who is a health care ethics consultant or clinical ethics consultant? There are no straightforward answers to such questions. Indeed, the attempt to answer such questions usually engenders controversies. Bioethics is a puzzle. Bioethics is itself a controversy, a theater of dispute. Across the world, there are persons who call themselves bioethicists. But there is no agreement as to what ends they are doing what they do, as to what they should be doing, or even as to what they are doing. In hospitals across the world, there are persons who are paid as clinical ethics consultants (aka health care ethics consultants) and who are often held to be engaged in helping resolve normative questions about health care decisions. But there is no agreement as to what norms they should engage. This is because there is real dispute about the content and character of both morality and bioethics. As a result, there is a puzzle as to how properly to characterize the nature of the normative questions posed to bioethicists, as well as the answers bioethicists give. Bioethicists are asked, for example, about when a particular medical intervention is inappropriate (or futile), about who should make life-or-death decisions, and concerning what information should be provided in order for a patient adequately to consent to treatment. The question is what kinds of norms and which ethics should frame such questions. In answering such questions, what norms and which norms should guide the answers? Are the norms at stake those established at law? Is the ethics about which health care ethics consultants (HCEC) give advice simply an account of relevant law and public policy, as well as of how law and public policy is customarily applied? Or are the norms moral norms? If so, norms of which morality? These puzzles about the nature of bioethics justify second thoughts about the entire endeavor of bioethics.

H.T. Engelhardt, Jr. (✉)
Department of Philosophy, MS-14, Rice University, Houston, TX 77005, USA
e-mail: htengelh@rice.edu

H.T. Engelhardt, Jr. (ed.), *Bioethics Critically Reconsidered*,
Philosophy and Medicine 100, DOI 10.1007/978-94-007-2244-6_1,
© Springer Science+Business Media B.V. 2012

If the norms at stake are moral norms, the problems proliferate. In terms of whose morality ought one to pose the questions and/or seek the answers? Do those who present questions to health care ethics consultants often have different genre of norms in mind than do the consultants? If those asking for an ethics consultation do not realize that they may have a different sense of ethics than do health care ethics consultants, what kind of disclosure should the consultants provide concerning the differences (e.g., "Let me make an important disclosure. One might think that my normative advice as a clinical ethicist is grounded in a universal secular ethics, but there is no such ethic. Instead, my ethics advice reflects the moral vision of one group of secular moralists. You, of course, may have quite different and incompatible moral commitments")? When providing ethics guidance, should health care ethics consultants indicate what goods and harms are most important in shaping their answers, as well as what strategies they hold will likely maximize the balance of benefits over harms? For that matter, are consultants, if they are to be moralists, to be consequentialists? And if so, what sort of consequentialists? By appeal to which list of benefits and harms, and in what relative weight or priority are they to give advice? Or should their concern be to indicate how not to violate wrongmaking conditions so as to act rightly, so as to be worthy of happiness? That is, should clinical ethicists indicate how to act in deontologically appropriate fashions? Or should they instead offer an integration of concerns about the good and the right? But then how should they balance the good and the right, especially since there is no agreement on these matters in general or regarding matters in moral philosophy in particular? Or yet further, is the cardinal moral goal that of a virtuous life in the context of health care so that clinical ethicists ought to guide those involved to a virtuous life and away from vice? But what then is virtue or vice? About all such normative issues, as well as moral-theoretical and meta-ethical issues, there is dispute, not agreement. Indeed, the history of moral disputes is as old as the project of moral philosophy itself.[1] Or, it may very well be that clinical ethicists rarely give normative advice. In any case, what kind of informed consent and disclosure should be given by clinical ethicists to those seeking an ethics consultation?

One confronts "a field", "a profession", that offers services, but where there is no agreement about the character or substance of the services offered. The chaos in bioethics is analogous to that in European medicine before it acquired a concern to correct for observer bias and achieved a scientific foundation in the contemporary sense of the notion. People may have agreed that patients consulted physicians because they wished to have their illnesses and disabilities cured, as well as their pains ameliorated, but there was great disagreement about the causes of the illnesses, disabilities, and pains, as well as about the nature of illnesses, disabilities, and pains and how to cure them. There was disagreement at both the level of theory and of practice. Bioethics as academic and practical fields are marked by a similar chaos. At both the normative and theoretical levels, there is dispute, not agreement. The level of disagreement may even be greater in bioethics than that which marked traditional Western medicine, because there may be greater unclarity about what the problems are that bioethics should address.

As will be shown by the essays in this volume, clinical ethics is not simply a "field" that engages a multiplicity of disciplines (as would be the case with a multi-disciplinary field), but it is a number of fields with different disciplines that have successfully coalesced into a marketable service package. As a result, there is no possibility of evidence-based, argument-based assessments of the field that do not beg cardinal questions. This is the case because in the face of moral pluralism there will be different standards of evidence and of sound rational moral argument. What is one to say about all this? Given this disarray, what legitimacy can bioethics, especially clinical ethics consultation, possibly possess? And why? It is the case that, at some level, bioethics seems to function as a publicly recognized social institution and/or cluster of practices. It is the case that bioethics as a "field" has its pundits who can be interviewed for moral, indeed bioethical sound bites. They make assertions such as: "That is morally outrageous!" "I have never heard of someone doing that." "That violates the established consensus." But what is the meaning of such assertions? One might conclude that the outrage expressed and various *obiter dicta* advanced by such pundits are really rhetorical ploys designed to bring others into agreement with the pundit's morality *cum* bioethics, his ideology. Bioethics has become influential, but the source of its influence and legitimacy is far from clear. Bioethics deserves a serious and critical re-assessment. This volume is a step in that direction.

1.2 Success in the Face of Foundational Disagreement

In the forty years since the term bioethics was first deployed in 1971 to identify a cluster of practices of giving guidance about biomedical morality and since the Center for Bioethics of the Kennedy Institute of Ethics of Georgetown University was brought into existence, bioethics has become widely accepted across the world. The growth has been dramatic. How ought one to understand this phenomenal development, especially in light of the circumstance that many of the crucial founding assumptions have simply proven false. How could bioethics have emerged and grown so quickly despite disputes about what bioethics actually is? In Beauchamp and Childress's account in *The Principles of Biomedical Ethics* (1979), a common human morality is presupposed, and therefore a common bioethics. Yet clearly there is no such common morality, or for that matter a common bioethics. Humans have fundamental disagreements regarding the content of morality and of bioethics. All human societies are marked by moral controversies that are reflected in political disagreements. Given the strident moral disputes within societies across the world and over history, the incredible claim that humans share a common morality is perhaps best understood as a special pleading on behalf of a particular morality. In any event, humans disagree as to when it is forbidden, obligatory, or merely licit to have sex, reproduce, transfer private property from unconsenting owners, and kill their fellow humans. There are disagreements as to when it is good or evil to tell the truth or to lie, whether homosexual acts are immoral, or whether capital punishment is to be celebrated or prohibited. Moral and bioethical pluralism reigns in the face of Beauchamp and Childress's claims on behalf of a common morality and a common bioethics. How, then, given that it is so unclear what morality and bioethics are, could bioethics have experienced the success it now enjoys?

Many appear simply to deny the problems confronting bioethics and are impressed by its seeming success. They have always hoped for a common morality, and they are committed to act in accord with their hope, facts of the matter to the contrary notwithstanding. Thus, in the face of deep moral disagreements, many bioethicists nevertheless talk about a moral consensus regarding cardinal moral and bioethical issues. The romance with consensus persists, often passionately, although there is likely no moral or bioethical issue regarding which all agree. Moreover, this lack of agreement is implicitly taken for granted in many countries, such as in the United States, where one has experienced the phenomenon of different presidents appointing different bioethics advisory groups with different understandings of what the canonical moral content of bioethics should be. For example, the moral commitments of President Bill Clinton's National Bioethics Advisory Commission were different from those embraced by President George W. Bush's President's Council on Bioethics. The connection between bioethics and politics became especially salient, when during the administration of President George W. Bush the President's Council on Bioethics and its chairman Leon Kass were the focus of criticism by those of different bioethical, political, and/or ideological persuasions. As in all policy issues with a heavy political and moral overlay, each seeks to choose the experts who will favor his approach to legal and political policy agendas. It is also clear why governments would want the blessing of a bioethics committee for the law and public policy they wish to establish. The governmental engagement of bioethics functions to convey a "moral" *imprimatur* for particular agendas in law and public policy. It is in addition clear why bioethicists would speak of consensus in the face of strong disagreement, for appeals to moral consensus can function as a camouflaged appeal for supporters to rally around a political agenda. There is a quite understandable aspiration to have the right bioethics so as to "bless" the right politics. Governments choose the morality and bioethics that will support their desired public policy. Does this mean that bioethics is primarily an element of biopolitics?

Bioethics has been engaged to support social movements. As Laurence McCullough notes, bioethics emerged inter alia as part of a social crusade on behalf of a particular understanding of autonomy and against a particular practice of medical paternalism. As McCullough also shows, although the medical paternalism decried may not have been frequent, it became a focus of major concern, which is to recognize that bioethics emerged as one of the many "rights" movements of the 1960s and 1970s. In this way, at the macro-level bioethics became, at least in part, biopolitics. At the micro-level bioethics became a resource engaged within hospitals and elsewhere to protect the alleged rights of patients despite secular "moral" disagreement regarding such rights. Of course, in doing so bioethics reflected a particular "moral" and a particular socio-political agenda directed against medical paternalism. In all of this, the question is whether bioethics as it took shape as a social movement was in the main a response to changes in the law bearing on consent to medical treatment that took place in the 1970s and 1980s. Or is bioethics best understood as having primarily arisen around a cluster of very particular moral and socio-political agendas that engaged bioethics to change law and public policy?

For example, did those with particular political commitments such as the Kennedy family engender bioethics in order to support their larger political goals?

In any event, at the macro-level bioethics has supported particular goals for governance. At the micro-level, one can say that as a fact of the matter, as a strategy for social cooperation, clinical bioethics has served as an instrument for conflict and risk management in hospitals. Given its apparent success in these areas, bioethics has become widely accepted as useful. Yet, at both the macro- and micro-levels, it is unclear as to what bioethics actually does or should do. These cardinal and unanswered questions return one to the issue of what patients, their families, physicians, and other health professionals should be told as a disclosure regarding the ethics that clinical ethicists advance. Before patients, their families, and/or health care providers consent to receive a clinical ethics consultation, what sort of information should they be given so that they will not be deceived regarding what bioethics in general and bioethics consultations in particular can offer? What sort of disclosure ought health care ethics consultants to make about bioethics in general and about the special commitments of the consultants in particular in order not to engender false expectations?

The problems with bioethics lie not just with false claims and false expectations about the character of morality and bioethics (e.g., "there is a common morality and bioethics and I am its spokesman"), as well as with one-sided accounts of the history of physician-patient relationships. The problems are far more fundamental. They involve questions about the nature of normative concerns themselves. The nature of secular morality and secular bioethics deserves a skeptical reappraisal because the very project of secular morality is itself questionable, or at least may not at all be what many thought it to be, or at least hoped that it would be. Many who were once religious may falsely presume that with regard to secular morality they can continue with business as usual even when a God's-eye perspective is denied or simply not recognized. They may expect direction in bioethics from a quasi-God's-eye perspective about what they ought to do. But "after God", no such perspective is available. The point is that reality and morality appear quite different without a final point of orientation and meaning.[2] It is not just that the project of morality has failed to establish a canonical morality or to vindicate a general theory of what a morality should be (i.e., to establish what morality is about), but that once morality is rendered secular, it does not have the force that was generally expected from morality in Western culture when morality was recognized as anchored in and enforced by God. Absent God, the appreciation of morality and metaphysical meaning, the phenomenological experience of reality and of morality, changes.[3] As Kant, likely an atheist,[4] recognized: without embracing the existence of God and immortality as postulates of practical reason, morality fragments and there are also no longer compelling grounds for moral obligations always trumping prudential considerations (Engelhardt, 2010c). If one embraces an atheistic methodological postulate so that all is viewed as coming from nowhere, going nowhere, and for no ultimate purpose,[5] then it also follows that morality and bioethics are multiple and ultimately meaningless.[6] Morality and bioethics cannot speak to humans generally as theologians might speak to members of a particular religion.

This leads to the question of reassessing the status of morality after God and after metaphysics. We have just begun to think through what bioethics can be "after God" (Engelhardt, 2010d), as well as what political authority means apart from reference to God. That is, we have just begun to ask what it means to talk of moral and political obligations if one acts as if everything, including morality, were ultimately meaningless. It would appear, for example, that the more-than-minimal state becomes simply a *modus vivendi* absent a God's-eye perspective. The state no longer functions with moral authority because there can in principle be no general agreement concerning which morality should frame its structure or give moral authority to the state. The more-than-minimal state becomes merely a political structure which one should out of concerns for prudence usually obey, but which possesses no general secular moral standing or authority (Engelhardt, 2010a, 2010b). Christians had accepted the authority of the state as a matter of divinely established obligation (see esp. Romans 13:1–4). But in a cosmos without final significance and in the face of moral pluralism, and given a state that acts in conflict with one's own moral and bioethical commitments, why morally ought one to obey the state, especially the more-than-minimal state, which acts without the consent of all its participants? The minimal state can at least indicate that all that support it consent to the state in the sense of agreeing to act together with the consent of all participants. That is, the minimum state is grounded in never using any citizen without his permission. But no state in the contemporary world is a minimal state, and the more-than-minimal state has no canonical secular moral justification because there is no canonical secular morality available to convey authority. Moreover, the claims of a social contract, if this means the authorization of its citizens, is at best specious once the state is a more-than-minimal state, a state through which certain groups (even if they are the majority) impose their will, even if it is through voting (should there be such), on all its citizens (while not responding to the pleas of citizens: "Love me, State, or leave me alone"). "After God," the significance of morality and political authority change radically.

As Vattimo wryly remarks regarding the wide-ranging implications of atheism, and indeed even of agnosticism, "God is dead, but man isn't doing so well himself" (Vattimo, 1991, p. 31). Which is to say that "the death of God, which is at once the culmination and conclusion of metaphysics, is also the crisis of humanism" (Vattimo, 1991, pp. 32–33). Without a God's-eye perspective, there is no longer a unitary meaning or significance to being human. Instead, there are multiple socio-historically-conditioned constructions of morality and of the significance of morality. Morality within the horizon of the finite and the immanent is both irresolvably multifocal and deflated in its meaning (e.g., it is unclear why moral concerns should always trump concerns for one's own self-interest). So, too, the significance of being human is also deflated with wide-ranging implications for claims of human dignity and human rights. The universality and strong normative force of claims of human rights and human dignity shipwreck on moral pluralism, as well as on the lack of any anchorage in being. It is for this reason that, as Vattimo acknowledges, atheism appears "as another catastrophic Tower of Babel" (Vattimo, 1991, p. 31). If there is no longer any final perspective in terms of which there could be one canonical morality or one bioethics, morality and bioethics become intractably plural. Moreover, and even more importantly, a point that Vattimo does not sufficiently underscore, morality and bioethics become ultimately meaningless. Within the horizon of the finite and the immanent, the significance of morality is radically reduced.

The puzzles are further underscored if one attempts, within the horizon of the finite and the immanent, to regard morality (and therefore bioethics) as a natural, socio-biological phenomenon. If, for example, one considers morality as part of the socio-biological adaptive strategies of humans, one would expect morality to differ in different environments. Different environments with different selective pressures will favor different moralities and, for that matter, bioethics. Also, specifying what will count as a successful adaptation always requires specifying the environment and the goals that determine success. Moreover, if morality is a cluster of strategies for maximizing inclusive fitness, one would presume that differing balances among different moral proclivities will maximize the survival of a reproductive group in different environments. Moreover, a pleomorphism of moral phenotypes and underlying genotypes producing a diversity of moral inclinations and intuitions would allow a wider range of responses to selection pressure and therefore an easier adaptive response to changes in the environment. Moral pleomorphism will likely advantage a group's ability to adapt and "succeed". In addition, one would expect that often strategic hypocrisy, not conventional moral behavior, will maximize reproductive fitness. For example, men who are kind and supportive of their male neighbors, while secretly having affairs with their neighbors' wives and impregnating them, will maximize their inclusive fitness (Symons, 1981). To judge whether this is good or bad, right or wrong requires that one possess a moral standard. The question is then which particular morality should be recognized as canonical or even guiding, since a particular normative morality cannot be read off from the socio-biological facts of the matter. Invoking a particular moral standard, however, once again underscores the problem of moral pluralism. Which standard ought one to engage and why?

The problems cut even yet more deeply. If morality as a natural phenomenon is understood simply to be a cluster of strategies that maximizes inclusive fitness, it may often turn out that what is conventionally termed immoral behavior will maximize inclusive fitness. For example, in a number of environments, the ethos of the Vikings, the "Viking morality", may maximize a group's inclusive fitness. That is, the Vikings who followed the pagan "divine advice" that "he should early rise, who another's property or life desires to have. Seldom a sluggish wolf gets prey, or a sleeping man victory" (Sturleson, 1906, "The High One's Lay" #58, p. 35) may in fact have increased their inclusive fitness. Through following these norms, the Vikings spread their genes across northern Europe. For the Vikings who affirmed the good of bravery and cunning in battle as well as the importance of a heroic death, the High One's advice was compatible with a form of the Golden Rule. Since the Vikings esteemed dying bravely in battle, they could accept doing to others in battle as the others would do to them in battle. The cardinal problem is then without a final point of reference: how does one show which considerations should have priority – moral considerations (and then which ones), considerations of personal advantage, or the pursuit of the maximization of inclusive fitness?

One is thus brought to reassess the significance of secular bioethics some four decades after its genesis. Hence this volume. All of the contributors are acquainted with both the academic and clinical dimensions of bioethics. Each in his own way indicates why one should rethink taken-for-granted moral and conceptual assumptions that engendered, shaped, and focused the predominant bioethics of the 1970s, 1980s, and 1990s. Each author provides a perspective on how bioethics has failed to achieve certain goals of its founders. The volume as a whole shows the failure to provide a canonical moral vision that could establish a normative intellectual framework able morally to authorize a liberal progressive legal and public policy framework (a latter-day Enlightenment hope that had been embraced by those who asserted the existence of a common morality and a common bioethics) or to deliver a clinical bioethics united around a single, morally normative task. Because bioethics cannot deliver what it was once taken to promise, namely, canonical moral direction able to guide health care and the biomedical sciences, and/or to provide canonical moral norms for the provision of morally appropriate health care to patients, the "field", if bioethics can be termed *a* field, is a conundrum. At the very least, society and the academy confront the task of reconsidering what bioethics is or could be, given the incoherent assumptions at its core. Again, this volume is offered as a step towards the critical re-assessment of bioethics.

1.3 The History of Bioethics: Four Perspectives

The first section of this volume offers a re-appraisal of bioethics through an examination of its historical roots. It does this through explorations provided by Michael Yesley, Griffin Trotter, Laurence McCullough, and Aaron Hinkley. The first essay shows the tie between the genesis of bioethics and the work of the National Commission for the Protection of Human Subjects of Biomedical and Behavioral Research. As Michael Yesley demonstrates, by constructing a moral framework for a public policy that could justify a body of regulations governing human research subjects, the Commission influentially contributed through its *Belmont Report* to creating the original "principles" for the practice of bioethics (National Commission, 1978). The Commission was able to articulate a moral framework that made its policy recommendations appear justified. The subsequent general acceptance of the policy recommendations made the moral commitments that lay at its roots seem vindicated. This circle of presuppositions and the political success of the process were taken to support the moral plausibility of the project of bioethics. The success of the Commission in this regard made the project of bioethics and the moral expertise it offered appear verified.

Nevertheless, there were crucial misperceptions at the start. In particular, as Yesley notes, analytic skills and a command of the geography of moral notions were

1 A Skeptical Reassessment of Bioethics

confused with moral expertise as such. Expertise in making moral distinctions, in identifying different genre of moral arguments, and in understanding the history of moral philosophy invokes capacities different from normative moral expertise, that is, in knowing what one should actually do. The person who is an expert about the first three matters is not necessarily, indeed likely rarely, a *phronemos*, an expert on how to act wisely, much less rightly or virtuously (whatever that might mean in the face of moral pluralism). There was in short an unwarranted generalization from analytic expertise to moral expertise. As Yesley puts it, "Many persons ... assumed that because philosophers are expert in analyzing ethical questions, they are also expert in answering those questions" (Yesley, 2011, p. 4). As he indicates, an open-ended, quasi-precedential, quasi-casuistic practice developed that tied the Belmont principles through procedural mechanisms to actual practice, thus driving the emergence of practices guiding and regulating human research. What actually proved to be functional were "modes of reasoning ... [that] establish the parameters of precedential examples, and examples [that] may assist interpretations of principles" (Yesley, 2011, p. 13). The nature of bioethics from its origins was more procedural than is often acknowledged. Yesley's insight into the genesis and life of bioethics led him to appreciate that the bioethics of research involving human subjects turns out to be "an empirical, investigational task, not merely an 'application' – whatever that signifies – of abstract rules and principles" (Yesley, 2011, p. 18). This essay by Yesley with his portrayal of the actual way in which bioethics functions shows not simply the need for a critical reassessment of the moral epistemology of bioethics, as well as of the proper concrete role of bioethics, but of the very nature of bioethics itself.

Bioethics was, as are most human endeavors, a child of its times. As Trotter in his essay appreciates, bioethics took shape under the pressure of vast social changes that during the 1960s and 1970s weakened and undermined many traditional social institutions (e.g., the family as a reproductive and social unit compassing a husband and his wife with their children). It was a period of world-wide upheaval and revolution. From Vatican II (1962–1965) to the Woodstock Festival (1969) to the anti-Vietnam War protests, the 1968 student unrest in Europe, the civil rights movements in the United States, and the Cultural Revolution in China (1966–1976), old certainties and social structures were brought into question. It was a time during which many looked forward to a radical reformation, if not revolution, in morality, as well as in social and political structures. On the part of many, there was a passionate commitment to a new, progressive, post-traditional, social-democratic, moral vision. There was a hope to establish a new political order. Bioethics from the beginning had an ideological overlay directed towards socio-political engagement of a particular genre. For example, the first generation of bioethicists was more likely to have supported George McGovern than Barry Goldwater. Many of the early bioethicists were former clergymen and/or former Christians now seeking commitment in more secular social causes. A good proportion could be characterized as theological dissidents and progressives before they took the turn to bioethics. Those who came to embrace bioethics regarded themselves more as critics and reformers of the status quo than as defenders of the Western tradition. They considered themselves to be agents of change rather than conservatives. It was the Kennedys and Shrivers who supported the Kennedy Institute of Bioethics, where the term bioethics was first applied to the field, not the Buckleys.

Bioethics emerged in a period that was anti-traditional in spirit. Accepting the values and institutions of the past because they are time-tested was widely regarded as inauthentic, as a form of self-alienation. Instead, as Trotter recognizes, one was to find oneself, realize oneself, and express oneself in the fullness of one's own individuality. This understanding of self-realization affirmed a view of self-determination that was compatible with being shaped by one's inclinations and passions, but not necessarily by either God or anonymous rational moral rules. It was a period that celebrated self-discovery more than self-discipline. The paradigm of autonomy was not the Kantian affirmation of rational self-determination, but one of generally unfettered self-direction. As Trotter points out, the governing adage was "Do your own thing, as long as it doesn't hurt anyone else." This adage was realized in various concrete forms such as in Timothy Leary's slogan.

> "Tune in, turn on, drop out," traces the social trajectory of the proposed psychological journey. He [Leary] also succinctly describes the destination and mode of transportation: "Blow the mind [through hallucinogens] and you are left with God and life – and life is sex" (Matusow, 1984, p. 290). As for rock 'n' roll, Leary claimed after hearing the Sergeant Pepper album that the Beatles were "evolutionary agents sent by God" (Matusow, 1984, p. 296).

Bioethics was engendered in an age when within the dominant culture it seemed natural to embrace a hermeneutic of suspicion regarding authority figures and traditional social institutions. This animus against established social norms and institutions was directed inter alia against physicians, the profession of medicine, and medical ethics. Times were changing, the privileges of physicians fell under suspicion, and bioethics aspired to recast the practice of the health care professions in accord with its own more progressive moral conceits.

In these circumstances, the claim made by physicians that their professional ethics was grounded in a special moral knowledge gained through a physician's clinical experience of the life-world of illness, suffering, and death could not be accepted. Such assertions of special moral expertise were rejected out of hand as elitist, paternalist, and unfounded in favor of the claims of moral expertise advanced by bioethicists. The agenda came then to be to relocate medical ethics within, if not replace it by, bioethics. The fashioning of a new patient-centered and physician-critical medical morality became the project of bioethics. Bioethics was regarded as the moral authority that should reshape medical ethics. It seemed only meet and proper that a bioethics framed by non-physicians should have priority over and recast the medical ethics of the past. That medical ethics was seen to reflect the special moral commitments and authority of medical professionals, so that a new ethics was thought to be needed to protect patient self-determination. Given the cardinal commitment to self-determination about which Trotter speaks, bioethics supported a revolution against the hegemony of physicians in health care. The goal was to replace the centrality of medical ethics with the centrality of bioethics. It was laymen, not physicians, who were held to be those who should properly determine the moral geography within which patients are treated. Medical ethics in the light of the aspirations of bioethics became viewed as parochial and self-seeking. The new health care ethics was to be free of the traditional, professional, physician-directed concerns of medical ethics and instead to affirm instead the self-determination and ethos of patients. The result was that medicine as a profession became more other-regulated (e.g., by patients, laymen, and government). Bioethics aspired to guide patients and public policy, as medicine was being reformed according to new, more liberal, more social-democratic norms.

Trotter's thesis is "that the hippie legacy, interpreted broadly lurks in bioethics' collective unconscious and manifests in dogmas, contradictions and totalizing aspirations that currently plague the field" (Trotter, 2011, p. 3). The passions of the 1960s and 1970s, now grown somewhat middle-aged, still influence the field. Among the driving passions at the genesis of bioethics was an Enlightenment aspiration to discover by reason a new and liberating truth, especially a moral vision that would be freeing and directing. The result is that "though bioethics tends to view itself as a bright light of reason, it has no compelling arguments for its basic dogmas" (Trotter, 2011, p. 22). In addition, given its ties to the engagements of an ideological movement, bioethics has a tendency, in lieu of coherent argument,

> to employ the language of moral revulsion. It would be interesting to count the number of times that bioethicists have tried to establish the high ground by casting moralistic aspersions on the United States for its failure to achieve what they consider to be an equitable distribution of healthcare. The lack of universal access has been described as "embarrassing" (Trotter, 2011, p. 22).

This state of affairs has produced a bioethics that has predominantly taken on the character of a liberal political movement – often with the political concerns of the 1960s. "Like other totalizing social movements, mainstream bioethics is animated by a utopian social vision and seeks to establish its vision universally, by police power if necessary" (Trotter, 2011, p. 25). Bioethics still aspires to political power through drawing on and engaging strong ideological passions camouflaged in claims supposedly grounded in moral-philosophical rationality on behalf of a supposed common morality.

The third essay explores the ways through which bioethics has been decisively shaped by a history of its founding that portrays bioethics as overcoming what was taken to be the special evil of medical paternalism. A very significant role was played by the widespread view that the medical profession prior to bioethics was robustly and improperly paternalistic, so as to justify the view that bioethics can celebrate the success of having liberated patients from the paternalism of physicians. Laurence B. McCullough brings this self-congratulatory account into question. He devotes special attention to reassessing a particular and widely influential study of medical paternalism that may have drawn on an unrepresentative sample.

> The correct reading of the frequently cited Oken (1961) study and of the two crucial sources in modern Western medical ethics (Gregory, 1772 [1998]; Percival, 1803 [1975]) is that physicians were not directly forthcoming with bad news because giving bad news was – and remains – one of the "most disagreeable" and "alarming" obligations of physicians one away from which they would naturally want to turn, just as Oken described. However,

patients were to be informed and, as a matter of obligation, physicians should see to it that seriously ill patients were informed, but by others than the physician – by their families in the case of wealthy patients being seen in their homes and by the house surgeon in the case of worthy-poor patients in the infirmaries (2011, p. 24).

McCullough's reassessment of this core justification for the emergence of bioethics and of its understanding of its early successes undermines the plausibility of a core element of the self-identity of the "field".

The origin of this view in bioethics, and the heroic image of bioethicists as reformers, even revolutionaries, was empirically and historically flawed. Its remarkable persistence means that bioethics continues to be defined and deformed by an ideology of anti-paternalism that it is long past time to abandon as the defining trope of bioethics (McCullough, 2011, p. 25).

Bioethics appears in great measure to have been early shaped by a crucial misperception of what was at stake in physician-patient relationships.

During the early emergence of bioethics, as McCullough notes, criticism of the so-called "Hippocratic ethos" also served as a focus for anti-paternalistic zeal. This criticism was often grounded in unfounded claims regarding the Hippocratic Oath and the so-called Hippocratic ethos.[7] The alleged Hippocratic ethos appears under careful scrutiny to have been more complex than usually portrayed (Edelstein, 1967). Moreover, it did not reach into the future as an organized and coherent tradition. What, then, is one in these circumstances to say about bioethics, which in great measure is a reaction against a one-sided portrayal of the ethos and the abuses of the past? Was bioethics the liberator of health care from medical paternalism, and if so, in exactly what way? Or did bioethics instead primarily work to erode medicine's sense of professionalism? Were bioethicists the defenders of patients who were in the grips of paternalistic physicians? Were bioethicists noble moral crusaders who should cover themselves with the mantle of being the special advocates of the civil rights of patients? Or is this account in part mythical? Or indeed, may bioethics to some extent be culpable for having undermined the virtues and self-identity that traditionally sustained the moral integrity of the profession of medicine? The more one examines the historical foundations of bioethics, the more complex those foundations appear, and the more difficult an account of bioethics becomes.

Aaron Hinkley's essay confronts a further foundational difficulty at the roots of bioethics: disparate accounts of the meaning and force of autonomy. Autonomy, which is often treated as a straightforward, unitary, and coherent moral principle, is not just ambiguous. In the actual "practice" of bioethics, an uncritical invocation of autonomy compasses a major tension between two competing understandings of autonomy. As Hinkley points out, "The concept of autonomy in Beauchamp and Childress' bioethical principlism is not merely unclear; it contains within itself two distinct understandings of what actually constitutes autonomy that are fundamentally contradictory" (Hinkley, 2011). At the core of Beauchamp and Childress's support for the principle of autonomy in their *Principles of Biomedical Ethics* (1979) is a conflation between a Kantian understanding of autonomy as *Wille*, as rationally ordered choice, and autonomy as *Willkür*, as decision-making that is inclination-driven. It is remarkable, as Aaron Hinkley points out, that Beauchamp

1 A Skeptical Reassessment of Bioethics

and Childress reference Kant in their explanation of their principle of autonomy without acknowledging the depth of this tension. Beauchamp and Childress note that

> Kant argued that respect for autonomy flows from the recognition that all persons have unconditional worth, each having the capacity to determine his or her own moral destiny. To violate a person's autonomy is to treat that person merely as a means; that is, in accordance with others' goals without regard to that person's own goals (Beauchamp and Childress, 2009, p. 109).

But for Kant, what is of cardinal importance is not the person's idiosyncratic goals. Instead, the issue is whether the goals that a person affirms are those that rational agents should choose. These two disparate views of autonomy underlie two quite different normative accounts of the moral life.

As Hinkley appreciates, a world of moral difference turns on this distinction between rationally-grounded choice and inclination-driven or even idiosyncratic choice, including whether a concern for autonomy, *pace* Beauchamp and Childress, is in fact supportive of a form of medical paternalism. After all, a Kantian physician committed to respecting (though perhaps tolerating) patients as autonomous moral agents is not obliged to respect the inclination-driven, heteronomous choices of patients. To the contrary, the Kantian physician is in general committed to inviting such patients to make truly autonomous, non-heteronomous choices.

> For Beauchamp and Childress, the physician ought to respect the freely made decisions of their patients regardless of the reasons, or lack thereof, that the patient has for making their choice. However, given the Kantian understanding of autonomy, all choices are not equal; not all choices qualify as autonomous. Respecting the autonomy of the patient, from a Kantian understanding, might well call for some form of medical paternalism where the physician encourages the patient to make the rational, autonomous choice (Hinkley, 2011, p. 93).

Of course, the Kantian physician should not coerce or lie to a patient. The question arises, however, whether a Kantian physician should in certain circumstances follow a professional standard of disclosure for informed consent. In particular, should a Kantian physician omit mentioning to patients inclined to unwarranted worries and other passions the infrequent side effects associated with a contemplated treatment because their mention would erode the patient's capacity for an autonomous choice? The Kantian account of what a reasonable and prudent person would require for rational choice would not necessarily be equivalent to that which would be required by the patient moved by inclination-framed concerns. At the roots of the emergence of bioethics there is, as Hinkley shows, an important tension grounded in incompatible understandings of autonomy with different implications for how one should regard medical paternalism.

That autonomy came for some to include heteronomous choices was the result of the confluence of two streams of cultural developments. The first cultural stream derived from the pursuit of self-realization in the sense of the satisfaction of one's own inclinations, a point made by Trotter in his reflections on the tie between bioethics and the self-indulgent American culture of the 1960s and 1970s. Autonomy in this case has the practical meaning of being unhindered in doing things one's own way, being free to be oneself, albeit inclination-driven (e.g., including being driven by unwarranted fears that the Kantian physician should seek to avoid). In this case, a moral agent in this anti-Kantian view of autonomy is properly self-determining, even when moved by inclinations and passions rather than only when acting on the basis of universalizable rational grounds, grounds compatible within the agent's authentic Kantian self. The second chain of cultural development involves a recognition of the circumstance that in the face of intractable moral pluralism permission serves as a default basis for collaboration and thus provides inter alia a framework grounded in individual authorization for collaboration (Engelhardt, 1986, 1996). One appeals to the will of persons, not to moral rationality, because moral rationality is plural. Even if one cannot agree on morally substantive matters, one can still for instrumental purposes agree to collaborate. Permission in this sense provides the foundation of the free market in which even moral enemies can with agreement collaborate. Autonomy in this instance is realized simply in the conveyance of permission, the conveyance of authority for collaboration, which may occur moved by inclination. These diverse clusters of concerns support individual decision-making, whether or not the decision-makers are autonomous in a Kantian sense.

1.4 The Practice of Bioethics and Clinical Ethics Consultation: Three Views

What, then, can one make of bioethics, given that bioethics is constituted out of, and functions as, a collection of diverse social concerns and commitments? This is the focus of the second section of this volume. It explores the consequences of the double pluralism of moralities (and therefore of bioethics) and of practices of bioethics consultation. As already observed, bioethics, rather than being a united field, much less a moral, social, and political movement, is riven by contrary moral visions, competing agendas, and incompatible self-understandings. That bioethics is marked by, even defined by the contentions and disagreements of the culture wars (Hunter, 1991) constitutes a disappointment for those who had hoped that bioethics would succeed as a powerful, well-focused, progressive, morally informed program for social and political action. Such an outcome would have been more plausible, had there been a common morality. The fact of moral and bioethical pluralism undercuts the aspiration to a canonical morality and bioethics that could anoint as canonical a unified socio-political movement to establish *the* rightly-ordered health care policy.

The conflict among moralities and bioethics has led to persistent disputes, not common collaboration in the pursuit of a progressivist agenda. Renée Fox and Judith Swazey have with regret noted the consequences of intra-bioethical disagreement.

The rancorousness has been conspicuous in statements made by bioethicists on both sides of the divide in professional publications, blog sites, and the media about matters such as embryonic stem cell research, therapeutic and reproductive cloning, genetic enhancement, other uses of fetal tissues and embryos, end-of-life issues, the place of religious thought and belief in ethical analysis, and about the potential dangers and harms as well as benefits that advances in medical science and technology can bring in their wake (2005, p. 369).

1 A Skeptical Reassessment of Bioethics

Mark Cherry in his article recognizes that what Fox and Swazey regard as an unfortunate but contingent characteristic of contemporary bioethics (namely, its being defined by disagreement) is in fact an essential characteristic of a "field" directed to moral truth in the face of intractable moral pluralism. Because a canonical, secular, moral truth (and bioethics) cannot be discerned, a political agenda is substituted, compounding the disputes engendered by moral pluralism with thinly veiled political struggles. The resulting state of affairs is characterized by competition among incompatible socio-political agendas in the guise of moral disputes. As already noted, bioethics as biopolitics has become a specialized field for the political struggle among health care ideologies, where each party seeks to use state power to establish its bioethics. Cherry illustrates this political struggle to establish at law a particular biomedical ethos or ethics through a study of the Convention on the Rights of the Child. As Cherry shows, the Convention reflects a very particular set of social-political agendas.

> If bioethics is to remove itself from the culture wars, if it is to capture a 'composed, fair-minded, rationally thoughtful outlook ideally associated with a philosophically oriented field' (Fox and Swazey, 2005, p. 369), it must think itself out of the box of its political ideology of human rights, welfare entitlements, and top-down governmental regulation driven towards a fully secular vision of ethics and social justice. ... It must cease to be an ideologically driven will to power, and recapture the social space necessary for both robust expression and instantiation of diverse moral, religious, and secular worldviews (Cherry, 2011, p. 22).

However, bioethics is inextricably bound to biopolitics and particularl political movements.

The character of bioethics is thus quite complex. On the one hand, it is possible for bioethics to be undertaken solely as the unapplied study of a particular area in the history and geography of ideas. As such, it can be an intellectual enterprise that is pursued for its own sake. However, once bioethics is brought into the public forum or even into the clinic so as to serve as more than a source for conceptual analyses and for outlines of the geographies of the diversity of moral claims, bioethics tends to be the defender of one versus other understandings of the moral and politically appropriate. Thus, for example, particular bioethics support particular visions of human dignity and/or human rights, which collide with other visions of human flourishing. Supporters of particular feminist visions of bioethics collide with supporters of particular traditional visions of the proper character of the family and of human relations. In all of this, bioethics is engaged to give a moral *nihil obstat* to particular ideological and political agendas. This is the case unless bioethics adopts a form of legal positivism in recognizing that, and affirming that, clinical ethics most often simply expounds what law and public policy require. In this case, bioethics becomes a vehicle to support a particular legally established ethos, so that clinical ethicists expounding that ethos can then have something normative about which to be experts. In the absence of a canonical morality, ethicists can at least be experts regarding whatever ethos is legally established.

Ana Iltis and Adrienne Carpenter in their essay challenge the assumption that bioethicists are members of a distinct profession. In so doing, they bring into question attempts to identify "bioethicists as professionals who share goals, and have specific socially recognized roles and functions for which they ought to be certified or at least bound by a code of ethics" (Iltis, 2011, p. 1). As they recognize, there is not simply moral bioethical pluralism, along with ideological and political pluralism. There is in addition role pluralism, a heterogeneity of understandings of what it is to be and act as a bioethicist. This heterogeneity plagues attempts to frame a unitary account of the field, much less a unitary professional identity. Given the reality of this complex pluralism, Iltis and Carpenter show the implausibility of the view that bioethicists "ought to work toward such a state in which bioethicists are professionals with a defined and robust ethos, shared goals and perhaps even advocate for particular moral positions together through their professional organization" (Iltis and Carpenter, 2011, p. 1). This is the case, so they argue, because stating that someone is a bioethicist identifies very little of substance about the person. In particular, identifying someone as a bioethicist conveys little information about what the bioethicist does, knows, or is committed to accomplishing, just as saying that some one is a teacher is highly uninformative, for the person may be "an elementary, middle, high school or pre-school teacher . . . [or] a private piano teacher or . . . homeschooler [or] perhaps teaches at the college level" (Iltis and Carpenter, 2011, p. 27). Similarly, a bioethicist

> might hold a Ph.D. in philosophy and teach at the undergraduate or graduate level; hold a Ph.D. in religious studies and teach medical students and residents; hold a J.D. and teach bioethics in a law school; be a nurse in a hospital who does clinical ethics consultations; be a physician who chairs an ethics committee and so on (Iltis and Carpenter, 2011, p. 28).

In short, to identify someone as a bioethicist does not convey information similar to identifying someone as a physician or a lawyer, much less as a neuro-surgeon or a patent lawyer.

Nor does the identification of someone as a bioethicist say anything substantive about that person's normative allegiances or commitments. In particular, learning that someone is a bioethicist communicates nothing of substance about the person's moral views regarding key issues contested in bioethics. The person may

> be pro-life or pro-choice, favor embryonic stem cell research or oppose it, support a right to physician-assisted suicide or not, believe that health care should be publicly funded through tax dollars or not, favor uncontrolled donation after cardiac death or not, believe we should adopt presumed consent statutes for organ donation or not, believe that parents should be allowed to enroll their children in research posing more than minimal risk with no prospect of direct benefit or not, and so on (Iltis and Carpenter, 2011, p. 28).

The ambiguity due to the heterogeneity of roles compassed by bioethicists is thus compounded by the diversity of positions and agendas bioethicists endorse. The result is that identifying someone as a bioethicist is like identifying a person as a member of a political party but without specifying which party. However, political parties such as the Republican, Democrat, Libertarian, and Maoist Communist, etc., are committed to different norms and often require different core competencies and differ even in what it is to be a political party. One might think of the special core competencies required of a good revolutionary Maoist communist party member. Because of disciplinary differences, functional diversity, the multiplicity of fields and areas of specialization along with the pluralism of salient religious, cultural, moral, and ideological agendas, the aspiration of bioethicists to professional unity is an *ignis fatuus*.

The question is then why there is such interest on the part of many, especially health care ethics consultants, to portray themselves as a member of *a* field united in *a* common profession, with a common set of core competencies and a common set of moral commitments, if not a common moral vision (e.g., there is often no recognition that clinical bioethics is a collage of fields, each with its own core competence, as with giving advice regarding established law, having facility in mediating disputes, etc.). Here the answer seems rather straightforward. Acting *as if* bioethics and clinical ethics were united as a field with a common professional vision, core competencies, and common moral commitments serves the economic and status interests of bioethicists, especially those of health care ethics consultants. It is difficult to market the services of a "field" frankly recognized as characterized by the salience of internal chaos, intractable disputes, and lack of common focus. The claim of ethicists to have ethics expertise is undermined by their field's being defined by disparate moral visions about which morality and in what way clinical ethicists have expertise. If one is selling "ethical" advice, one's "profession" will likely appear more attractive if it appears to be clear as to the ethics package that is being purveyed. To invite persons to conceive of bioethics or at least clinical ethics consultation as a profession is at least to aspire to a coalition of disparate undertakings being united in order reciprocally to aid each other in marketing the services of clinical ethics. In short, banding together and terming themselves a profession will likely enhance the social prestige of a group of service providers. There are also incentives to control entry into a service group so as to advantage those who are already members of that group through protecting their economic interests in the service market against those who might seek to compete with those already established in the field. Thus, there are economically good reasons to act *as if* bioethics possessed a conceptual and/or professional integrity, because this will likely enhance the status of bioethicists, even if there is no conceptual or professional integrity.

In my essay "Why Clinical Bioethics so Rarely Gives Normative Moral Guidance: Some Critical Reflections," I explore why bioethics has been so successful despite intractable moral pluralism. The abrupt appearance, growth, and spread of bioethics, as has already been noted (Engelhardt, 2002), was driven by the secularization of the West and the deprofessionalization of medicine, which occurrences created a moral vacuum into which bioethicists successfully entered. Given the deprofessionalization of medicine and the anti-traditional trends of the time, physicians were not regarded as plausible moral guides. Given the forces of secularization, priests, rabbis, and ministers had been marginalized from taking the position of moral guides for society and for social institutions. Bioethicists asserted that they could fill the void and provide the desired secular moral guidance. As a result, just as a major socio-economic niche opened, bioethicists were able to enter and find takers for their guidance regarding the proper conduct of medicine. That is, bioethics was so widely and rapidly accepted because, just as medicine was becoming increasingly successful and expensive, and as the biomedical sciences were raising new and engaging moral puzzles that seemed to threaten novel possibilities that required moral assessment, traditional societal moral experts and pundits (e.g., theologians, priests, rabbis, ministers, and experienced physicians) had been marginalized from acting as guidance providers for the public. In this context, many did not notice or want to acknowledge the moral pluralism that undermined the promises of bioethicists to provide definitive moral guidance. Despite intractable moral pluralism, it was hoped that bioethicists would give canonical secular moral direction.

Bioethicists could not furnish the secular moral guidance for which many hoped, because there was no one canonical secular morality able to guide those giving guidance on the model of how advice can be given within traditional religious communities with a well-established orthodoxy. Nevertheless, the market demand for guidance remained. Society and health care institutions wanted guidance, and were willing to pay for it. Bioethicists for their part generally did not emphasize the difficulties besetting bioethics. After all, it is for many unsettling frankly to recognize the extent and depth of moral pluralism. The success of bioethics is best accounted for in terms of bioethics having brought together a collage of disparate services that created an attractive service package that proved to be a generally marketable commodity, despite moral and bioethical pluralism. Various forms of moral disorientation and concerns for risk management created a market demand for the complex of services health care ethics consultants purported to be able to offer. That is, the combination of heterogeneous services offered by bioethics turned out to meet successfully a set of felt needs. In institutions such as hospitals, the composite attractiveness of the bioethics service package has justified the place of clinical ethicists as useful jacks of many trades.

The complex concatenation of services that bioethicists offer through its market success also supports the social status of bioethics, even though there is bioethical and role pluralism. On the one hand, some bioethicists have achieved an academic standing in being able to:

(1) analyze ideas, concepts, and claims cardinal to the various understandings of moral issues of concern to health care and the biomedical sciences,
(2) assess the soundness of arguments bearing on these issues, and
(3) provide conceptual geographies of different moral and moral-philosophical positions regarding bioethical issues by showing how, given different initial premises and rules of evidence, one can establish different moral norms and viewpoints; thus, for example, bioethicists can show that act-utilitarians of a certain sort will affirm certain courses of action as appropriate choices, while showing that Kantians will embrace other choices (Engelhardt, 2011, p. 163).

The intellectual prestige that such bioethicists have secured in the academy is then often transferred to the "field" as a whole and to the status of clinical bioethicists in particular. Health care ethics consultants or clinical ethicists for their part successfully offer a wide diversity of different services. Secular clinical ethicists

1 A Skeptical Reassessment of Bioethics

(1) act as legal advisers who, since they are not lawyers admitted to the bar, can provide relatively inexpensive, on-the-spot legal advice about the ethos established at law and through public policy (e.g., on Institutional Review Boards they can provide information about relevant regulations and about how to apply that ethos);

(2) act as attorneys do in defending a particular interpretation of applicable law, in that there will always be grey zones of the ethos established at law and by public policy (as a result, ethicists for hire have no more of a conflict of interest than do attorneys when ethicists are hired to serve as ethics consultants for a corporation so as to defend a particular interpretation of grey zones in the legally established ethos; there is no conflict of interest in health care ethics consultants functioning as ethicists for hire [as it would not be improper for lawyers] because their behavior is in accord with the nature of secular bioethics and secular ethics consultation);

(3) act as legal liability risk managers who help physicians and hospitals to lower the risk of malpractice actions through such practices as documenting the grounds for controversial clinical decisions and by their presence showing that there has been due diligence in considering what was at stake in problematic clinical cases;

(4) act as dispute mediators and communication specialists who aid parties such as physicians, patients, family members, and nurses better to collaborate in providing medical treatment, all of which is accomplished with the authority or imprimatur of being able to

(5) act as ethics experts (i.e., anointed with the authority of bioethics as an academic moral-intellectual discipline), who possess special analytic knowledge and expertise about how to analyze what could be considered to be obligations in particular circumstances, given particular moral commitments and moral-theoretical assumptions (all of which is not normative advice), although actual normative moral guidance when given (e.g., as a Roman Catholic ethicist in a Roman Catholic hospital or as an advocate of a particular secular sect of morality and bioethics) is always one among many competing particular moral accounts or moral narratives (Engelhardt, 2011, p. 23).

The result is that bioethics flourishes, although there is no intellectual, moral, ideological, or service integrity or unity to the multiple "fields" that are treated *as if* they were *the* field of bioethics.

To put the matter in a more provocative fashion: the "field" of clinical ethics (aka health care ethics consultation) as a collage of services succeeds precisely because it has no foundational integrity that might constrain its freedom to compass and sell a plurality of heterogeneous services. In particular, the persistent attractiveness of clinical ethics consultation is a function of the market-framed bundle of services provided. What is important is that the service providers offer a package of services, some of which services the market will usually buy.

> The astonishing success of bioethics shows how a service profession can function as a well-adapted social organism by combining numerous clinical services while protecting itself with an intellectual camouflage that allows for a successful adaptation in different market environments. A bundle of different skills, roles, and facilities has been successfully brought together and given cultural standing by being nominally grounded in an academically accepted field (Engelhardt, 2011, p. 166).

Bioethics flourishes although its foundations are incoherent because it offers a wide range of services, thus appealing to a wide range of customers. To take a metaphor from business, bioethics is a successful conglomerate.

1.5 The Incredible Search for Bioethical Professionalism: Some Final Critical Reflections on Circular Thinking

This volume closes with a slice of the history of the struggle of clinical bioethicists both to define themselves as professionals and to achieve recognition for their "profession" through articulating a set of core competencies. Jeffrey Bishop's essay examines a cardinal dimension of the search on the part of health care ethics consultants for a professional self-identity. The difficulty is that the identity is sought in terms of a commitment to the pursuit of defining goals, which goals they hold themselves competent to assess, rather than recognizing that the "unity" of the "profession" of health care ethics consultants lies in a dynamic collage defined by a market demand for a shifting cluster of services from a heterogeneous group of service providers. Pace the American Society for Bioethics and Humanities, given the difficulties involved in identifying under one rubric bioethics as a "field", bioethicists face a considerable challenge in defining the conceptual integrity of their endeavor (ASBH, 2011). Given secular moral pluralism, the world of moral concerns is rendered intractably ambiguous. Given a plurality of roles, the service nature of bioethicists is also ambiguous. Bishop's study of the attempt to enhance the profession of bioethics through establishing a set of competencies for health care ethics consultants completes this volume's reflections on the protean character of bioethics.

As Bishop shows, it is impossible to clarify the goals of bioethics in a fashion that would produce clear canons for the evaluation of professional competency, because moral pluralism supports a plurality of standards and role pluralism establishes diverse bioethical undertakings, making impossible a clear and precise statement of core commitments that is not vacuous. One cannot locate canonical standards. This state of affairs radically undermines the search for core competencies.

> This document refers to goals, but it does not clarify what the goods of HCEC are. What goods do the HCE consultant aim at such that HCE consultants can claim to be a bona fide practice? The Core Competencies draft document states:
> While all health care providers engage in ethical decision-making as part of their everyday practice (e.g. in facilitating informed consent with a patient or family before a procedure), health care ethics (HCE) consultants (or "ethics consultants") differ from other health care providers in that they have been assigned by their institutions the distinctive role of responding to specific ethical concerns and questions that arise in the delivery of health care, and therefore require a distinctive set of competencies to effectively perform this role. (Core Competencies, 4)
> Several important points come into relief here. Whatever the goods of HCEC might be, they are derivative of the institutions of medical practice, or the hospital institution that supports them. At least, this is what the *Core Competencies* seem to be claiming. Yet, HCE consultants don't think of themselves as upholding the goods of medicine or of the goods of their employers. So, they must be upholding the values and goods of the patient. Are they not then best thought of as patient advocates? And if they are patient advocates in the face of the overwhelming powers of the health care and medical institutions (with their separate goods), would the goods of health for a particular patient not be best served by lawyers? And is there no conflict of interest, given that the vast majority of HCE consultants are employed by hospitals?.

1 A Skeptical Reassessment of Bioethics

Given the diverse concerns that created and sustained the field of bioethics in general and the field of health care ethics consultations in particular, the articulation of a substantive list of core competencies, at least beyond relevant knowledge of local laws and regulations, is an *ignis fatuus*. In this endeavor, an important truth tends to be obscured: the field is attractive in great measure because its ambiguities and protean character allow it to adapt to and sell its services within diverse markets.

Given the core ambiguities in bioethics rooted in moral and role pluralism, and given the drive to anoint the field of clinical ethics consultation with core competencies, the temptation is to assess health care ethics consultants in terms of whether, according to standards established by these very consultants, consultants do what consultants say they do. Such a circular approach can be self-vindicating, but nevertheless, or perhaps for that very reason, socio-politically useful. Bishop notes:

> First, there is an assumption that the quality improvement assessment assists the clinical ethics consultant by systematically evaluating his procedures and standards. The basis for that assumption is that the circularity of the process leads to better definitions, more narrow definitions, definitions that pick out certain relevant features of the clinical ethics consult. The second and most important aspect of this "funneling process" or "epistemological circuit" shows that as these definitions ossify into standards of practice, they become more and more the ready-made rules of engagement. The third aspect of the process, then, is that the circularity of the process would come to structure the practice of the expert. The expert designs the instrument; and the instrument assures the expert that he is in fact expert.

As a consequence of the pursuit of professional identity through professional self-vindication, one is led to embrace an epistemological circle, such that one comes to see what one intends to know. One comes to hold that one knows what one assumes is there to be known, not necessarily what actually exists in reality. One has, as it were, a self-certifying and self-vindicating process floating in the air of professional aspirations.

"Ethics" in the context of contemporary health care thus becomes the activity of an ethics bureaucrat who self-certifies his success as an ethics bureaucrat. The result is that when one attempts to measure quality improvement (QI), one avoids recognizing how value pluralism brings quality pluralism. As a consequence, in the absence of a canonical standard for quality, quality and quality improvement must be stipulated within a bureaucratic framework that selects particular goals and therefore can measure the efficiency and effectiveness of activities aimed towards achieving those goals. But whether those goals are worth pursuing, or indeed whether the goals themselves are so strategically ambiguous so as to be primarily misleading, remain as challenging questions.

> [A]s Alasdair MacIntyre has noted, [in such circumstances] the highest goods are the goods of efficiency and effectiveness. The ethicist then is a bureaucratic manager, guiding a process toward unknown and interchangeable ends and goods. (MacIntyre, 1984, pp. 26–27, 30–32, 74–78, 85–87) In HCEC, the supreme good is efficiency and effectiveness of the process, the intermediary goals not the goods of a valued practice.

One encounters with health care ethics consultation an expression of a general problem with secular morality. In a culture after God and after metaphysics, intractable moral pluralism undermines the capacity to identify common goals.

1.6 Bioethicists for Hire: A Concluding Exploration

After all of this, one has to ask, is it still legitimate to call health care ethics consultants health care ethics consultants? Is there any ethics about which such health care ethics consultants might have an expertise that would justify characterizing them as health care ethics consultants? In my essay I recognize the legitimacy in health care ethics consultants still calling themselves ethicists, in that they are experts regarding the *Sittlichkeit* (to invoke a Hegelian insight, which contrasts ethics with morality) established at law. The "ethics" about which health care ethics consultants are experts is not morality as such, but a particular ethics, a particular ethos that has been *aufgehoben* and given force through law and public policy. After all, one of the principal services supplied by clinical ethicists is advice about law and public policy bearing on health care, as well as advice about how to comply with that law and established policy. Of course, complying with particular established laws and public policies may be immoral, depending on the law or policy. That is, there may be moral grounds in certain instances for health care ethics consultants being obliged to declare to those who are about to receive advice or consultation, "You are required by law and public policy to do X. However, in my considered moral judgment grounded in my morality (which morality is generally characterized as _____), which morality may not be your morality, that if you follow the established law and public policy, this would be immoral."

Once one recognizes that the ethics which is the primary concern of health care ethics consultants is not grounded in a set of moral norms but is rather in a gloss on a particular ethics or ethos established at law and in public policy, one can appreciate why the advice of health care ethics consultants would be sought after, and be regarded as useful, even in the face of moral pluralism. Health care ethics consultants in giving "ethical" advice in fact provide a restatement of applicable law and public policy framed in a moral discourse that is grounded in a genre of legal positivism that accepts local law. In any particular legal jurisdiction, the requirements of established law and public policy will be fairly clear, despite fundamental moral disagreements, although there will be grey zones in which ethicists can argue a case on behalf of a particular account of the established ethos. It is for this reason, inter alia, that pharmaceutical firms engage bioethics consultants. Bioethicists in such circumstances provide advice not only regarding the texture of the established ethos of the culture, but also about how to navigate through the established ethics so as to achieve certain goals without incurring legal or social sanctions. In this case, hiring a clinical ethicist or bioethicist is like hiring an attorney. The ethicist hired is no more engaged in a conflict of interest in such a circumstance than would an attorney hired to defend the particular perspective of a company. There is always some ambiguity as to what the law requires or the established ethos demands. Just as it is appropriate for an attorney to explore the ambiguities of law and ethics that are to the advantage of those whom the attorney is hired to represent, so, too, a corporation can properly hire ethicists to plead and advance the corporation's interests. In a secular pluralist culture, this involves no conflict of interest. Nor is serving as an ethicist for hire counter-cultural, as it would be in a totalitarian culture where both the enactment of law and the establishment of a legally recognized ethos, as well as its interpretation, is closely controlled by the establishment. Lawyers function differently in North Korea than in the United States, as would ethicists. Perhaps some who oppose ethicists for hire think of ethicists as being experts about a clear fact of the matter, such as unapplied scientists. However, clinical ethics provides a form of easily accessible and financially reasonably affordable legal advice combined with other skills such as dispute mediation, risk management, and communication, along with the capacity to supply analyses of the character of moral puzzles, as well as of the geographies of the diversity of normative commitments, which may be of use to a wide variety of institutions. Health care ethics consultants can contribute to all of this without giving morally normative guidance, or in fact in certain circumstances by serving as an advocate for hire. Different buyers in the market will at different times have greater or less interest in one service rather than another offered by ethicists.

Finally, and importantly, it will generally be advantageous for a health care institution to hire an ethicist not only so as to have its own house ethics pundit who may be able to speak on behalf of and defend the institution, but because the presence of an ethicist serves important public-relations goals. In general, providing "ethics" services shows that an institution is morally sensitive and appreciative of the moral questions raised in the provision of health care. Having an ethicist demonstrates that the institution is committed to helping parents, families, and health care providers in facing the questions of sexuality, reproduction, suffering, dying, and death raised by health care – even though no actual morally normative guidance is given. For those who are more discerning, the presence of a house ethicist available to patients and families, or at least to health care providers, communicates the health care institution's commitment to comport with the locally legally-established ethos. The same is the case when a pharmaceutical firm hires its own house ethicist or secular ethics consultant. The public affirmation of virtuous commitments is usually a good profit-maximizing strategy (or for non-profit health care institutions, the affirmation of virtue generally serves as a good strategy for maintaining a positive budget margin; if there is no margin in the budget, there is no mission; Engelhardt and Rie, 1992).

Bioethics has turned out to be quite different from how it was widely regarded at its inception. In a period of social and cultural transition, canonical moral guidance and direction had been sought. Given residual Enlightenment expectations regarding moral philosophy, combined with a faith in natural-law reasoning (the Kennedy Institute's Center for Bioethics was lodged in a Catholic university, Georgetown, with doctrinal commitments to the capacities of natural-law reflection), it seemed plausible to many that bioethics could supply the moral guidance that many had once sought from religion. It was assumed that bioethicists could on the basis of a common morality that should be knowable by all proceed by means of right reason to disclose the requirements of a common bioethics. In a culture under rapid secularization, clinical ethicists entered to provide, among other things, services as secular chaplains (Engelhardt, 2002). Moreover, various parties with ideological and political agendas endorse bioethics in the hope that *their* bioethicists can give moral authority to their particular political movements which support particular visions

24 H.T. Engelhardt, Jr.

of social justice, human rights, human dignity, feminism, and patients' rights. It was not in the interests of these advocates to recognize that their views represented the position of one among a diversity of secular moral sects and socio-political movements. Moreover, confronting moral pluralism along with the deflation of the authority of morality and the moral authority of the state is far from reassuring. It is deeply disquieting. We have just begun honestly to assess the character of secular morality, secular bioethics, and how to understand the secular culture they warrant.[8]

Notes

1. Once one is blind to transcendence, as Protagoras (ca. 490–420 B.C.) realized, "Man is the measure of all things, of things that are that they are, and of things that are not that they are not." Diogenes Laertius (2000, vol. 2, pp. 463, 465, IX.51). Protagoras lived in a period when traditional Greek, religious, and social norms were under attack, in disarray, and in transformation. Many groups, but especially physicians and Sophists, played important roles in secularizing Greek society. See Versenyi (1963). This led to a complex reappraisal of the project of morality as the new reigning moral ethos of the Hellenistic era emerged (Guthrie, 1977). In reaction, some such as Plato (428–348 B.C.) and Aristotle (384–322 B.C.) attempted through philosophical reflection to establish a canonical morality. Others recognized the impossibility of such projects grounded in philosophy. As Clement of Alexandria stressed (A.D. c. 150–211). "Should one say that Knowledge is founded on demonstration by a process of reasoning, let him hear that first principles are incapable of demonstration; for they are known neither by art nor sagacity" (Clement, The Stromata, Book 2, chapter IV, 1994, p. 350). By the 3rd century, the possibility of a philosophically-justified morality was acknowledged in many circles as discredited. For example, Diogenes Laertius (3rd century after Christ) provides the following summary of the five tropoi, the five general ways in which such philosophical reflections fail through sound rational argument to come to conclusions that would be needed for establishing a canonical secular morality. These had been summarized by Agrippa. "Agrippa and his school [affirm five] modes, resulting respectively from disagreement, extension *ad infinitum*, relativity, hypothesis and reciprocal inference" (Diogenes Laertius, 2000, vol. 2, p. 501, IX.88). Sextus Empiricus (also 3rd century) also summarizes these five modes: "the first based on discrepancy, the second on regress *ad infinitum*, the third on relativity, the fourth on hypothesis, the fifth on circular reasoning" (Sextus Empiricus, 1976, vol. 1, p. 95, I.164. The failure of the moral-philosophical project was also acknowledged by St. John Chrysostom (A.D. c. 349–407), who reflected critically on the accomplishments of the Greek philosophers. Regarding those philosophers, St. John opined,

   > it hath been made manifest by themselves, that an evil spirit, and some cruel demon at war with our race, a foe to modesty, and an enemy to good order, oversetting all things, hath made his voice be heard in their soul. ... For what could be more ridiculous than that "republic," in which, besides what I have mentioned, the philosopher, when he hath spent lines without number, that he may be able to shew what justice is, hath over and above this prolixity filled his discourse with much indistinctness? This, even if it did contain anything profitable, must needs be very useless for the life of man (Chrysostom, 1994, Homily on the Gospel of Matthew I.10, 11, vol. 10, p. 5).

   St. John in critically assessing the moral-philosophical project of the pagan Greek philosophers knew that the only escape from the intractable plurality of moral visions described by Protagoras and Agrippa as well as by the early Christian Fathers lies in a noetic faculty, a way of knowing truth directly, so that the knower can come into union with the known. He did not hold that all could come to know natural law simply by philosophical reflection. As he says in his commentary on the second chapter of St. Paul's letter to the Romans:

1 A Skeptical Reassessment of Bioethics

> But by Greeks he [St. Paul] here [Rom 2:10–16] means not that that worshipped idols, but that that adored God, that obeyed the law of nature, that strictly kept all things, save the Jewish observances, which contribute to piety, such as were Melchizedek and his, such as was Job, such as were the Ninevites, such as was Cornelius (Chrysostom, 1994, Homily V on Romans I.28, V.10, vol. 11, p. 363).

If canonical moral knowledge is to be possible, there must be a form of immediate knowing, somewhat on analogy with a first-person experience of the quale blue. This knowing must involve a knowledge beyond the first-person experience of sensible firstness, a mere sensible quale. For more detailed reflection on these matters, see Romanides (2002) and Engelhardt (2006). See also Engelhardt (2000, chapter 4).

2. Outside of the monotheism of the God of Abraham and prior to the attempt to articulate a God's-eye perspective in Greek philosophy, morality appeared as the mores and the customs of a particular people. Even Aristotle understood virtue as that which is possessed by a gentleman with the proper upbringing of an Athenian. Before Christ, there were 613 laws for the Jews and 7 laws for the sons of Noah. With the Christians and the neo-Platonists, a strong appreciation of a universal morality becomes salient, though for Christians it is grounded in the God Who commands.

3. The role and significance of an appeal to a God's-eye perspective need not be religious. As a philosophical not religious matter, the significance of such an appeal to God turns on an appreciation of what is required to talk coherently, in principle, of there being one canonical morality, of moral reasons trumping concerns for personal advantage, and of morality not being ultimately meaningless. Appeals to God were made by philosophers ranging from Aristotle (384–322 B.C.), René Descartes (A.D. 1596–1650), Benedict Spinoza (A.D. 1632–1677), and Gottfried Leibnitz (A.D. 1646–1716), to Immanuel Kant (A.D. 1724–1804) independently of religious concerns. These philosophers referenced God in their philosophical writings not as a matter of proper worship or religious recognition of God, but through an acknowledgement that as a "philosophical-scientific fact of the matter" the recognition of God was crucial for the meaning of morality and reality. These philosophers in different ways appreciated the necessary role played by a God's-eye perspective in developing a coherent account of reality and/or morality.

4. Kant engaged the idea of God for purely philosophical reasons. As Manfred Kuehn puts it, "Kant did not really believe in God" (Kuehn, 2001, pp. 391–92).

5. By the atheistic methodological postulate I mean the commitment in most of the dominant secular cultures of Western Europe and the Americas to approaching reality as if there were no God when entering into public discourse, and especially during reflections in the public forum.

6. Already in the 19th century, the dramatic shift from a traditional Christian culture to a post-Christian post-traditional culture had begun to recast the moral understandings shaping the dominant culture. This profound change was a focus, for example, of Dostoevsky in his novel *The Possessed*. As Dostoevsky realized, these changes had deep roots in Western European culture, going back at least to the Donation of Pepin (A.D. 754 and 756). These political and theological changes lie at the roots of the emergence of Roman Catholicism as a distinct denomination at the beginning of the second millennium. These developments, mediated through Roman Catholicism's faith in reason, shaped its account of faith and reason, *fides et ratio*. This philosophical synthesis served as a major root of contemporary secularism, as both Dostoevsky (see, for example, "The Grand Inquisitor" in *The Brothers Karamazov*, book 5, chapter 5) and later Vattimo, among many others (see, e.g., Buckley, 1987), recognize.

> The West is secularized Christianity and nothing else. In other words, if we want to talk about the West, Europe, modernity – which, in my argument are held to be synonymous – as recognizable and clearly defined historical-cultural entities, the only notion we can use is precisely that of the secularization of the Judeo-Christian heritage (Vattimo, 2002, p. 73).

The culmination of these developments began to transform Western culture as it entered the 19th century. A. N. Wilson notes the scope of the change:

26

> As Dostoevsky made so clear in that terrible prophecy, and as Thomas Hardy and Leslie Stephen and Morrison Swift would probably all in their different ways have agreed, the 19th century had created a climate for itself – philosophical, politico-sociological, literary, artistic, personal – in which God had become unknowable, His voice inaudible against the din of machines and the atonal banshee of the emerging egomania called The Modern. The cohesive social force which organized religion had once provided was broken up. The nature of society itself, urban, industrialized, materialistic, was the background for the godlessness which philosophy and science did not so much discover as ratify (Wilson, 1999, p. 12).

7. Laurence McCullough, for example, points out that many have expressed an animus against medical paternalism that may not be fully justified by the historical record. In particular, some of the criticism of medical paternalism by bioethicists involves a criticism of the so-called "'Hippocratic' footnote reading of the history of Western medical ethics, which has been thoroughly debunked" (McCullough, p. 8). See also Baker (1993).
8. It is far from clear what Western culture will be like when a majority of its members have been raised in the embrace of a culture set fully within the horizon of the finite and the immanent, so that most of the members act as if morality were ultimately meaningless. Currently, most members of Western cultures were at least raised by persons whose grandparents were raised within Christendom, that is, within cultures in which Christian norms and metaphysical commitments were established at law. Into the mid-20th century, the United States was a Christian nation or, as the Supreme Court once opined, "we are a Christian people." Church of the Holy Trinity v. United States, 143 US 457 (1892). See also United States v. Macintosh, 283 US 605 (1931). In the West, not only is the content of law in great measure still drawn from Christianity, but the sense and meaning of morality as well, even though that morality cannot be justified in general secular terms. We have yet to experience a robustly post-religious, post-metaphysical culture in which Christian norms were not learned early, at least as a second language.

References

American Society for Bioethics and Humanities. 2011. *Core competencies for healthcare ethics consultation, 2nd ed*. Glenview, IL: American Society for Bioethics and Humanities.

Baker, R.B. 1993. The eighteenth-century philosophical background. In *The codification of medical morality: Historical and philosophical studies of the formalization of Western medical morality in the eighteenth and nineteenth centuries: Volume one: medical ethics and etiquette in the eighteenth century*, eds. R.B. Baker, D. Porter, and R. Porter, 93–98. Dordrecht: Kluwer.

Beauchamp, T., and J. Childress. 2009. *The principles of biomedical ethics*, 6th ed. New York: Oxford University Press.

Buckley, M. 1987. *At the origins of modern atheism*. New Haven, CT: Yale University Press.

Cherry, M.J. 2011. Bioethics as political ideology. In *Bioethics in the plural: Ideology and social construction*, 99–122. Dordrecht: Springer.

Clement of Alexandria, St. 1994. In *Ante-Nicene fathers*, vol. 2, eds. Alexander Roberts and James Donaldson. Peabody, MA: Hendrickson Publishers.

Diogenes Laertius. 2000. *Lives of eminent philosophers* (trans: Hicks, R.D). Cambridge, MA: Harvard University Press.

Edelstein, L. 1967. In *Ancient medicine*, eds. O. Temkin and C.L. Temkin. Baltimore, MD: Johns Hopkins University Press.

Engelhardt, H.T., Jr. 1986. *The foundations of bioethics*. New York: Oxford University Press.

Engelhardt, H.T., Jr. 1996. *The foundations of bioethics*, 2nd ed. New York: Oxford University Press.

Engelhardt, H.T., Jr. 2002. The ordination of bioethicists as secular moral experts. *Social Philosophy & Policy* 19.2(Summer): 59–82.

1 A Skeptical Reassessment of Bioethics

Engelhardt, H.T., Jr. 2006. Critical reflections on theology's handmaid: Why the role of philosophy in Orthodox Christianity is so different. *Philosophy & Theology* 18.1: 53–75.

Engelhardt, H.T., Jr. 2010a. On the secular state. *Politeia* 26.97: 59–79.

Engelhardt, H.T., Jr. 2010b. Political authority in the face of moral pluralism: Further reflections on the non-fundamentalist state. *Politeia* 26.97: 91–99.

Engelhardt, H.T., Jr. 2010c. Moral obligation after the death of God: Critical reflections on concerns from Immanuel Kant, G.W.F. Hegel, and Elizabeth Anscombe. *Social Philosophy & Policy* 27.2(Summer): 317–340.

Engelhardt, H.T., Jr. 2010d. Kant, Hegel, and Habermas: Reflections on "Glauben und Wissen". *The Review of Metaphysics* 63(June): 871–903.

Engelhardt, H.T., Jr. 2011. Why clinical bioethics so rarely gives normative moral guidance: Some critical reflections. In *Bioethics critically reconsidered: Having second thoughts*, 151–174. Dordrecht: Springer.

Engelhardt, H.T., Jr., and Michael A. Rie. 1992. Selling virtue: Ethics as a profit maximizing strategy in health care delivery. *Journal of Health & Social Policy* 4.1: 27–35.

Fox, R.C., and J.P. Swazey. 2005. Examining American bioethics: Its problems and prospects. *Cambridge Quarterly of Healthcare Ethics* 14: 361–373.

Gregory, John. 1772 [1998]. *Lectures on the duties and qualifications of a physician*. Edinburgh: W. Strahem & T. Cadell. Reprinted in McCullough, L.B., ed. *John Gregory's writings on medical ethics and philosophy of medicine,* 161–248. Dordrecht: Kluwer.

Guthrie, W.K.C. 1977. *The sophists*. New York: Cambridge University Press.

Hinkley, A.E. 2011. Two rival understandings of autonomy: Paternalism and bioethical principlism. In *Bioethics critically reconsidered: Having second thoughts*, 85–95. Dordrecht: Springer.

Hunter, J.D. 1991. *Culture wars: The struggle to define America*. New York: Basic Books.

Iltis, A.S., and A. Carpenter. 2011. The "s" in bioethics: Past, present and future. In *Bioethics critically reconsidered: Having second thoughts*, 123–149. Dordrecht: Springer.

John Chrysostom, St. 1994a. Homilies on the Gospel of Saint Matthew. In *Nicene and post-nicene fathers first series,* vol. 10, ed. Philip Schaff. Peabody, MA: Hendrickson Publishers.

John Chrysostom, St. 1994b. *Homilies on the acts of the apostles and the epistle to the romans.* In *Nicene and Post-nicene fathers first series,* vol. 11, ed. Philip Schaff. Peabody, MA: Hendrickson Publishers.

Kuehn, M. 2001. *Kant: A biography*. Cambridge: Cambridge University Press.

MacIntyre, A. 1984. *After virtue*, 2nd ed. Notre Dame, IN: University of Notre Dame Press.

Matusow, A.J. 1984. *The unraveling of America: A history of liberalism in the 1960s*. New York: Harper and Row.

McCullough, L.B. 2011. 'Bioethics' ideology of anti-paternalism. In *Bioethics critically reconsidered: Having second thoughts*, 71–84. Dordrecht: Springer.

National Commission for the Protection of Human Subjects of Biomedical and Behavioral Research 1978. *The Belmont Report*. Washington, DC: U.S. Government Printing Office; DHEW [OS] 78-0012.

Oken, D. 1961. What to tell cancer patients. A study of medical attitudes. *Journal of the American Medical Association* 175: 1120–1128.

Percival, T. 1803[1975]. *Medical ethics: A code of institutes and precepts, adapted to the professional conduct of physicians and surgeons*. London: J. Johnson & R. Bickerstaff. Reprinted by C.R. Burns, ed. Huntington, NY: Robert E. Krieger Publishing.

Romanides, J. 2002. *The ancestral sin* (trans: Gabriel, George S.). Ridgewood, NJ: Zephyr Publishing.

Sextus Empiricus. 1976. Outlines of pyrrhonism. In *Sextus empiricus* (trans: Bury, R.G). Cambridge, MA: Harvard University Press.

Sturleson, S. 1906. *The elder eddas* (trans: Thorpe, Benjamin). London: Norroena Society.

Symons, D. 1981. *The evolution of human sexuality*. New York: Oxford University Press.

Trotter, G. 2011. Genesis of a totalizing ideology: Bioethics' inner hippie. In *Bioethics critically reconsidered: Having second thoughts*, 49–69. Dordrecht: Springer.

Vattimo, G. 1991. *The end of modernity* (trans: Snyder, Jon R.). Baltimore, MD: Johns Hopkins.

Vattimo, G. 2002. *After Christianity* (trans: D'Isanto, Luca). New York: Columbia University Press.

Versenyi, L. 1963. *Socratic humanism*. New Haven, NJ: Yale University Press.

Wilson, A.N. 1999. *God's funeral*. New York: W.W. Norton.

Yesley, M.S. 2011. Beginning bioethics. In *Bioethics critically reconsidered: Having second thoughts*, 31–47. Dordrecht: Springer.

Part I
History of Bioethics: Four Perspectives

Chapter 2
Beginning Bioethics

Michael S. Yesley

2.1 History

One day many years ago a former boss called to tell me about a job opening for "someone like you." A federal advisory commission on human experimentation was about to hold its first meeting and needed a staff director. I was intrigued by the opportunity but dubious. Protecting human research subjects would be a stretch, career-wise. Since law school I had worked briefly in civil rights but mainly in financial regulation. And I wasn't sure what "someone like you" meant.

Still, the new commission would address interesting human rights issues that had received much attention in the press. Also, my undergraduate major in philosophy might prove useful at last. It was, after all, an ethics commission, assigned to identify "basic ethical principles." And the commission's work might draw on my regulatory experience, but in a more interesting area with more interesting colleagues. I accepted the position and spent the next four years immersed in the inaugural effort to develop national bioethics policy.

The field called bioethics had emerged in the 1960s in response to a perceived need for expert, impartial advice and governance concerning new, or newly recognized, issues in biomedicine. These issues implicated both scientific and broader considerations – social, political, legal, economic, ethical, etc. – that required more comprehensive and independent management than had traditionally been provided by the physicians and scientists whose activities raised the issues. Bioethics engaged outsiders – "strangers" in one historian's description (Rothman, 1991) – in analysis and decision making. Fortunately for my career, there were few such outsiders in 1974, and a newcomer could assume a central position in the field without too close scrutiny of qualifications.

In the early 1960s dialysis gave rise to an early instance of bioethics in operation, if not in name. The physicians who developed renal dialysis turned to a multidisciplinary group to select the recipients of the procedure when there were insufficient

---

M.S. Yesley (✉)
Independent Consultant, Bainbridge Island, WA 98110, USA
e-mail: myesley@earthlink.net

H.T. Engelhardt, Jr. (ed.), *Bioethics Critically Reconsidered,*
Philosophy and Medicine 100, DOI 10.1007/978-94-007-2244-6_2,
© Springer Science+Business Media B.V. 2012

machines to provide it to all who needed it. The physicians recognized that selecting recipients, which determined who would live, involved broader considerations than medicine alone, and physicians were the wrong persons to make this decision. (Alexander, 1962)

Other biomedical advances also raised extra-scientific issues. The prospect of organ transplantation and genetic engineering led Senator Walter Mondale in 1968 to propose the establishment of a commission to study the ethical, legal, social and political implications of these anticipated advances. At first Mondale's proposal failed when many prominent physicians and scientists testified they could resolve any issues themselves, without public meddling. (Rothman, 1991; Jonsen, 1998) But when Mondale reintroduced his proposal in 1971, a New York Times article noted "a growing consensus that the consequences of science are too important to be left to the scientists." (Brody and Fiske, 1971)

Human experimentation also raised concern. During the 1960s and early 1970s, several press reports described human studies conducted in the U.S. without informed consent, in the shadow of the Nuremberg trials that disclosed the involuntary use of concentration camp prisoners in medical experiments. The U.S. cases included the injection of live cancer cells into indigent elderly patients at Jewish Chronic Disease Hospital, the injection of hepatitis serum into patients at Willowbrook State School for the Retarded, a placebo trial of contraceptive pills that produced 11 unwanted pregnancies in San Antonio, and the infamous Tuskegee study of untreated syphilis. Henry Beecher, a Harvard anesthesiologist, documented a widespread problem of unethical clinical research. (Beecher, 1966) In a review of the issues ("Doctors Must Experiment on Humans But What Are the Patient's Rights?"), the New York Times Magazine suggested that "elaborate codes . . . cannot settle everything." (Goodman, 1967)

In 1966 the Public Health Service adopted a policy requiring (in its final version) review of human experimentation by an "appropriate" committee at any institution where the PHS funded such research, to determine that "the rights and welfare of the subjects are adequately protected," "risks . . . are outweighed by the potential benefits to [the subject] or by the importance of the knowledge to be gained," and "informed consent of subjects will be obtained by methods that are adequate and appropriate." (Department of Health, Education, and Welfare, 1971) This policy was a step in the right direction but lacked the force of law and provided only a cursory description of subjects' rights. More problematic, the PHS policy assigned responsibility for protecting those rights to an institutional committee that could be composed entirely of a researcher's peers. Similar committees had approved some of the human research projects that raised public concern.

Senator Ted Kennedy believed broader control was needed. The Senate Health Subcommittee, which he chaired, held hearings in 1973 on research that had involved human subjects without their informed consent, including the Tuskegee Syphilis Study and research with prisoners, children, and the institutionalized mentally infirm. (U.S. Congress, Committee on Labor and Public Welfare, Subcommittee on Health, 1973) The public impact of these hearings was augmented by reports of a perfusion study of the severed heads of aborted fetuses. (U.S. Congress, 1973)

The groundwork was laid, and in the following year Congress passed the National Research Act (U.S. Congress, 1974), which established the National Commission for the Protection of Human Subjects of Biomedical and Behavioral Research, with a mandate to identify the basic ethical principles for human research and to recommend guidelines to protect research subjects with reduced capacity to consent. The Act also required that HEW-funded human research be reviewed by an Institutional Review Board and imposed a moratorium on HEW-funded fetal research until the commission (within four months) recommended conditions under which it would be appropriate. The HEW Secretary was required to implement the commission's recommendations or explain why not. In addition, the commission was directed to conduct the study first proposed by Senator Mondale.

2.2 Method

The commission began work on its most pressing order of business, fetal research, in December 1974. The commission's approach to this and most other topics in its mandate was to learn about the nature and extent of the research in question and the range of ethical opinion about it, then to consider this information, deliberate, reach conclusions, and make recommendations for regulating the research. The staff oversaw the gathering of information requested by the commission, then assisted in analyzing the information and developing conclusions and recommendations. The commission made a strong effort to be empirical. Kenneth Ryan, the chairman, asserted at an early meeting that the commission would be "scientific," that is, it would base its conclusions on factual evidence.

As a lawyer I was familiar with this process of assembling and considering the relevant evidence before reaching conclusions. Also, the commission's approach was not markedly different from policy analysis that I conducted at The RAND Corporation, which was the next stop in my peripatetic career. The commission's practice shared with policy analysis an emphasis on empiricism and a focus on practical considerations bounded by political and social concerns. True, the commission's conclusions were often clad in the language of ethics or accompanied by rationale framed in ethical terms, but the invocation of ethical generalities seemed (to me, at least) a rhetorical justification of largely pragmatic decisions rather than an application of abstract, determinative principles.

2.3 Philosophy

Following my lawyerly concern with sources of authority, I wondered about the nature of philosophy's contribution to the commission's work. My undergraduate study of philosophy in the late 1950s had generated an impression that philosophers were more interested in metaethical conjecture about the nature of ethics than directly confronting what is ethical. I recalled taking an ethics course devoted to linguistic analysis, with little mention of actual ethical issues. So I was skeptical of the role of philosophers (even those "saved" by medicine) in dealing with the issues before the commission. In fact, philosophy and philosophers were ubiquitous in the commission's work and featured in its reports, but I questioned whether their contribution was uniquely or specifically philosophical and, in any event, whether it mattered a great deal.

Although the commission engaged dozens of philosophers to write papers setting forth their views on issues before the commission, and two distinguished philosophers (Stephen Toulmin and Tom Beauchamp) served on our multi-disciplinary staff, the precise nature of the philosophical contribution is questionable. Did philosophical analysis ineluctably determine the commission's results, or provide guidance without dictating the precise results, or assure that relevant questions were addressed without indicating what the answers should be, or merely supply rationale and explanations for decisions made elsewhere in the commission's cogitation and deliberative process? Did philosophers play a unique role in achieving the commission's results, or did they analyze issues and information from the same practical perspective as members of other professions? Did philosophers merely add credibility and intellectual heft to a process that did not require their input to achieve its results? Were the commission's recommendations philosophical, or were they practical policymaking accompanied by philosophical commentary?

Many persons interested in the sort of issues before the commission have assumed that because philosophers are expert in analyzing ethical questions, they are also expert in answering those questions. Indeed, the philosophers who assisted the commission were expert in analyzing questions and providing logical and coherent rationales for their own and the commission's answers. But when the challenge was to evaluate whether – or under what conditions – particular acts or classes of acts were ethical, as opposed to clarifying the issues and establishing general rules or ex post rationales, the philosophers lost their privileged position and were put to the same practical tests as anyone arguing for one conclusion or another. Consideration of factual evidence and likely consequences was crucial, more than any abstract speculation, in developing policy recommendations.

If the philosophers' input was not determinative of the commission's largely pragmatic conclusions, they provided valuable assistance in supporting the commission's recommendations. And the philosophers' function was not limited to providing rationale: they joined the natural and social scientists, lawyers, physicians and other non-philosophers in raising questions and considerations that the commission took into account in developing its recommendations. But this last aspect of the philosophers' role was shared with the non-philosophers, reflecting the variety of empirical and other inputs and consideration of consequences incorporated in the commission's deliberations.

It is difficult to tease apart the respective influences of the various elements of the commission's four-year process of analysis and deliberation. However, it is clear (at least to this participant-observer) that the commission members viewed their function as comprised of (1) learning how human subjects in each population under scrutiny were selected and involved in research, what the research accomplished, and at what cost or risk; (2) considering the range of views of scientists, philosophers, lawyers and others about the appropriateness of research with each population; and (3) based on this information, recommending requirements and conditions for conducting human research consistent with the protection of the subjects. Philosophers did not play a special role in fact-gathering, the first of these efforts. In the consideration of opinions, the philosophers' role was prominent but shared with those who provided and analyzed opinions from the perspectives of other disciplines. Finally, in the development of the commission's recommendations, the philosophers provided philosophical justification for determinations that the commission members based primarily on practical considerations. Only when the commission generalized its results in the Belmont Report, a nonspecific policy statement, did the philosophers play a dominant role.

A brief review of some of the commission's recommendations and its Belmont Report will demonstrate this evaluation of the philosophers' – and philosophy's – contribution.

2.4 Fetal Research

The commission contracted for a comprehensive review of the scientific literature, which found that nearly all fetal research in the previous decade was therapeutic and/or presented minimal or no risk to the fetus, and very little research was conducted on the nonviable "fetus ex utero." (Mahoney, 1975) Thus, before the commission even commenced its deliberations, the literature review substantially shrank the area of contention about fetal research. For the great bulk of such research that was not problematic, the commission unanimously recommended conditions that were already recognized, e.g., institutional review, informed parental consent, prior animal studies, and intention to develop important knowledge not obtainable by alternative means.

But what to do about the remaining small but contentious categories of research, i.e., nontherapeutic fetal research during or in anticipation of abortion, or on a nonviable "fetus ex utero"? Unwilling either to recommend an absolute ban or, alternatively, to ignore moral concern about this research, the commission (with one dissent) recommended additional conditions, of which the most significant was a procedural requirement that such research must be approved by a "national ethical review body." (National Commission for the Protection of Human Subjects, 1975)

Thus, the commission kicked the can down the street, first to a short-lived Ethics Advisory Board and then to oblivion, since HEW disbanded the EAB in 1980 and never appointed a successor body that could approve the most contentious categories of fetal research. Still, the commission's recommendations were a measured response that disaggregated a variety of fetal research activities and suggested controls appropriate to each category. The commission's reliance on a procedural mechanism to deal with the most problematic categories recognized the impossibility of anticipating details that might militate in favor of, or against, permitting particular studies in those categories.

In preparation for its deliberations on fetal research, the commission also obtained position papers by eight philosophers and theologians, spread across the ideological spectrum, and Stephen Toulmin prepared a meta-analysis of these papers. Notwithstanding the varied perspectives of the philosophers and theologians (one of whom was reputed never to have approved, and another never to have disapproved, a human research project), Toulmin reported that "the doctrinal commitments and philosophical standpoints of the different participants in this discussion turn out ... to have had less influence than may have been expected on the practical recommendations they are prepared to support." Toulmin identified a "substantial moderate consensus" that cut across doctrinal lines and "stops short of an outright ban, [but] advocates a system of social controls carefully designed to limit the scope, and prevent the abuse, of fetal experimentation." (Toulmin, 1975)

The blurring of doctrinal differences identified by Toulmin is an example of the greater strength of practical over doctrinal considerations in policy making and may explain why the eight discussants had limited influence on the commission. Although the discussants reflected doctrinal differences, when they focused on the practical issues at hand, these differences were not dispositive, according to Toulmin. Similarly, although the members of the commission reflected different perspectives, they achieved near consensus on practical recommendations.

The commission's role was not to settle doctrinal questions but to provide a practical means to balance the values of scientific inquiry and individual integrity in the context of a range of procedures in fetal research. Essentially, the commission (1) demonstrated that a significant conflict between these values arises in relatively few studies; (2) provided a useful inventory of conditions for institutional approval of the great bulk of fetal research, in which there is no significant conflict of values; and (3) proposed an expanded, national review for the small amount of research that does raise a values conflict. In sum, the commission's accomplishment regarding the regulation of fetal research primarily involved clarification and process, rather than standard-setting, and the role of philosophy was corroborative.

2.5 Research Involving Prisoners

In the midst of reports exposing problematic human studies in medical settings, writer Jessica Mitford drew attention to human research in prisons (Mitford, 1973). Mitford testified at the Kennedy hearings that drug testing in prisons, then a common practice in the U.S., was exploitative, sometimes dangerous, and not adequately regulated (U.S. Congress, Committee on Labor and Public Welfare, Subcommittee on Health, 1973). Other journalists buttressed Mitford's testimony, and Congress included research involving prisoners on the list of issues for the commission to tackle.

The commission obtained information through surveys of the nature and extent of research in prisons, site visits to prisons where research was conducted, public hearings, and papers presenting social science, legal, and philosophical perspectives. The most vexing issue for the commission was whether prisoners can consent to participate in low-risk research, such as phase 1 drug testing, with "a sufficiently high degree of voluntariness" (in the commission's words echoing the Nuremberg Code). Is the prison environment so inherently coercive that prisoners cannot meaningfully consent to participate even in research that poses little risk? Also, is it fair to impose the burden of participation in drug testing on prisoners?

The commission's consultants came down on both sides of the issue of voluntariness. Legal consultants (Annas et al.) concluded that the law will recognize a prisoner's consent if there are safeguards to ensure that participation in research is not the only way for prisoners to maintain their health, hygiene and minimally decent living conditions, and that parole or a sentence reduction is not offered in return for participation. To the same effect, sociologists (Irwin, Susman) concluded that prisoners can have the freedom to give informed consent if they are insulated from staff and peer pressure, and information about their participation in research is not leaked to parole boards. Two philosophers came to a different conclusion, arguing that nontherapeutic research should not be conducted in prisons because there is reasonable doubt prisoners can give sufficiently free informed consent under present conditions (Branson) or because offering research participation to prisoners is bribery (West). A third philosopher concluded less categorically that "the strong ethical objection can be met realistically only by the most assiduous responsibility to minimize the coercive and exploitative element" in the prison (Wartofsky). (National Commission for the Protection of Human Subjects, 1976b)

To learn first-hand about research with prisoners, members of the commission visited several prisons, including Jackson (Michigan), then the largest prison in the U.S. and once the largest walled prison in the world, with over 55 acres enclosed by a concrete wall 34 feet high. This configuration had prompted a short-lived helicopter escape a year before the commission's visit. Among the cellblocks inside Jackson, two companies had constructed free-standing facilities to conduct phase 1 drug testing. The commission's tour of Jackson included cellblocks, the drug testing facilities, and shops for manufacturing furniture and license plates.

The commission talked with a representative sample of 80 prisoners and a group of inmates whom other prisoners referred to our accompanying consultant, an armed-robber-turned-sociologist. The prisoners generally expressed the feeling that

> they were free to volunteer for or withdraw from the [drug testing] program at will and were given adequate information about research protocols. . . . They generally rejected the notion that they were coerced into participating in research, and stated that they knew their participation would not be revealed to the parole board. . . . [Their] strongest objection was that the pay for participation in research was held down to levels comparable to prison industries.

The commission concluded that "prisoners currently participating in research consider, in nearly all instances, that they do so voluntarily and want the research to continue." (National Commission for the Protection of Human Subjects, 1976a)

The commission deliberated at length the pros and cons of low-risk research involving prisoners, particularly phase 1 drug testing. The relevant ethical principles were readily identified, but ambiguities complicated their application. The commission noted that respect for persons requires honoring an individual's choice but also protecting the individual from forces that might compel that choice. Also, justice requires that a class of individuals not bear a disproportionate share of the burden, but also not be unfairly excluded from the benefits, of participating in research.

The evidence, much of it provided by the prisoners themselves, was that the participants in testing considered their involvement to be voluntary, they valued the limited remuneration and had alternatives to earn equivalent pay in the prison industries, the risks of testing were small, and participation in drug testing did not, and was not expected to, affect parole. On the other hand, the drug companies had no need to involve prisoners in the testing aside from convenience. The commission reached a paternalistic conclusion that the prisoners' freedom to volunteer was "compromised" and the burdens of participation in research should be equitably distributed no matter how small they were. Thus, the commission determined that the mere presence of any compromise and burden, even if small, was unacceptable.

The commission did not recommend outright prohibition of phase 1 drug testing and other low-risk research in prisons but, in a measure that proved impossible to implement, that it be conditioned on satisfaction of several requirements, including findings by a national ethical review body that the reasons for involving prisoners in the research are compelling, the prisoners' involvement is equitable, and the conduct of the research is characterized by a high degree of voluntariness on the part of the prisoners and openness on the part of the prison, as demonstrated by compliance with several standards. (National Commission for the Protection of Human Subjects, 1976a)

Even if a national ethical review body had undertaken this task of accreditation, the findings necessary to permit phase 1 drug testing in prisons were improbable. Neither the drug companies' convenience nor the prisoners' option to earn modest remuneration is a "compelling" reason to involve prisoners, when the testing does not require their involvement or provide a crucial benefit to them. Further, although the commission's standards for voluntariness and openness were reasonable goals of prison reform, they were unlikely to be satisfied before drug companies turned to other testing populations.

In any event, the American Correctional Association, which was the only body competent to determine if the commission's requirements were satisfied, decided not to fully accredit any prison that permitted medical research. Unable to implement the commission's recommendation, HEW adopted regulations in 1978 barring non-therapeutic research in prisons unless it related specifically to prisons or prisoners. After the Food and Drug Administration (FDA) published similar rules in 1980, four prisoners at Jackson sued to prevent enforcement of the rules, which they claimed would violate their right to participate in research. Although the FDA agreed in a settlement not to implement the rules, drug companies had already abandoned prison research, having discovered that students and poor people [were] especially viable alternative populations from which to draw participants for nontherapeutic experiments – if the cash rewards were sufficient. The growing controversy surrounding the use of prisoners as research subjects, combined with the realization that they could find enough alternative subjects for their needs, led drug companies to make decisions that were based not so much on ethics as expediency (Advisory Committee on Human Radiation Experiments, 1996).

Further, when the FDA liberalized the use of foreign data beginning in the mid-1990s, the drug companies exported much of their testing abroad, where costs were substantially lower and rules more difficult to enforce.

In the name of voluntariness, then, the commission overruled the prisoners' assertions that they chose freely to participate in drug testing. On the ground that the prisoners had few options to earn money, the commission effectively removed one of those options. To justify its decision, the commission invoked a presumption that prisoners would not participate in testing if they were in better circumstances. This is a defensible presumption, but it applies equally to the students, poor people, and inhabitants of developing countries who have taken the prisoners' place as subjects of phase 1 testing.

The commission's rationalization of its standards for drug testing in prisons demonstrates the indeterminacy of principles and the limits and potential arbitrariness of applying them. As the commission was aware, the principles of respect for persons and justice are laden with ambiguities: whether to honor the prisoners' expressed choice or to protect them from their constrained choice, and whether eliminating a small burden justifies terminating a small benefit. The resolution of these ambiguities should involve a close weighing of factual detail, but the commission's consideration of these issues was abstract, focusing more on the concepts of "inherent coercion" and "equitable distribution of the burdens of research no matter how large or small" than the actual experience of the prisoners. If the commission had concentrated more on the level or degree of coercion and risk, the same principles could have supported more attainable, less stringent conditions for conducting phase 1 drug testing in prisons.

The principles construed by the commission (and enshrined in the bioethics pantheon – see Belmont discussion infra) did not dictate the commission's resolution of the issues raised by phase 1 drug testing in prisons. Rather, the principles were sufficiently broad to have supported either the commission's result that invalidated phase 1 testing in prisons or a different conclusion that voluntariness, although compromised in the prison setting, might still be adequate for consent to participate in research presenting the low risk associated with phase 1 testing.

A few weeks after the commission issued its report on research with prisoners, I was invited to speak about the report at several prisons in upstate New York. Although research was not conducted at these prisons, I was told the inmates were interested in any topic related to the subject of drugs. My talks provided an opportunity to explore how prisoners with no previous knowledge of phase 1 drug testing might view the issues confronted by the commission. At each prison, I presented the arguments for and against drug testing on prisoners, then led a discussion about informed consent in prison, followed by a show of hands. The prisoners were indeed interested in the subject and expressed opinions ranging from "there is no free will in prison" to "if you treat us like children not able to make up our own minds, we'll act like children." Their show of hands after each discussion indicated a split of opinion about the validity of consent in prison.

2.6 Research Involving Children

When the commission turned from prisoners to children as research subjects, the focus of debate shifted. The commission's characterization of prisoners' consent, not the level of risk to which they might be exposed, shaped the recommendation that effectively ended drug testing in prisons. In the case of children, who are incompetent to consent as a matter of law, the commission might have focused on the scope of parental authority to consent on their behalf – a preoccupation of most philosophers who wrote on this subject. Instead, the commission concentrated on the risks and benefits of research involving children. The sticking point was nontherapeutic research presenting more than minimal risk. A commission majority recommended permitting a modest increment of risk but encountered a snag in stating its rationale.

Employing its customary taxonomic approach, the commission classified research with children by level of risk and intent (therapeutic or nontherapeutic). Therapeutic research in general and nontherapeutic research presenting no more than minimal risk did not raise difficult issues, and the commission unanimously adopted recommendations establishing conditions for research in these categories. A majority of the commission also concluded that nontherapeutic research presenting a "minor increase over minimal risk" might be justified under strict conditions, and a recommendation permitting this category of research with IRB approval was adopted over the dissent of two members. A recommendation permitting nontherapeutic research exceeding a minor increase over minimal risk with approval by a national ethical review body was also adopted unanimously.

The disputed recommendation (which HEW implemented) permits nontherapeutic research presenting more than minimal risk to children, provided an IRB finds (i) the risk is only a minor increase over minimal risk, (ii) the research is similar to procedures ordinarily experienced by children with the subjects' condition, and (iii) the research is likely to yield vital knowledge about that condition. One dissenter (Cooke) stated he would permit such research if it were approved by a national ethical review body. (National Commission for the Protection of Human Subjects, 1977)

To their consternation (shared by the staff), the commission majority could not agree on a statement supporting their recommendations. Several drafts were rejected. Finally, three commissioners approved one statement, four commissioners approved another, and two commissioners approved both statements. (National Commission for the Protection of Human Subjects, 1977)

The issuance of two supporting statements suggests there was a moral disagreement, but the statements offer essentially the same rationale for the recommendations, including the disputed one. The difference is mainly structural: one statement

2 Beginning Bioethics   41

is organized by discussions of the principles (beneficence, respect for persons, justice) with references to the recommendations; the other statement is organized by discussions of the recommendations with references to the (same) principles. Both statements assert that the potential benefits of nontherapeutic research to children with the participants' condition (if not to the participants themselves) can justify a minor increase over minimal risk and the parents' consent to their children's participation in the research. Cooke, the dissenter who would require approval by a national ethical review body for the disputed category, lumped the majority's statements together, noting that in justifying its recommendation for this category, the commission "can invoke only the principle of utility [which] does indicate the perilous nature of the recommendation and the ethical uncertainty of the commission." (National Commission for the Protection of Human Subjects, 1977)

Two philosophers associated with the commission commented later on the purported disagreement about the rationale for the recommendations despite the commissioners' agreement about the recommendations themselves. In an influential book that sparked a long-running dispute between so-called principlists and casuists, Jonsen (commissioner) and Toulmin (staff) asserted that when commissioners

> explained their individual reasons for participating in the collective recommendations ... differences of background that lay dormant during the case-by-case discussions sprang back to life. The Catholic members of the commission gave different reasons for agreeing from the Protestants, the Jewish members from the atheists, and so on. Even when, as a collective, the commission agreed about particular practical judgments, the individual commissioners justified their readiness to join in that consensus by appealing to different 'general principles.'

In a footnote, Jonsen and Toulmin stated that the commissioners agreed about the principles of autonomy, justice and beneficence "on a completely general level" but "these shared notions were too comprehensive and general to underwrite specific moral positions." As support for their championing of casuistry over principles as the basis for ethical reasoning, Jonsen and Toulmin asserted that

> [t]he locus of certitude in the commissioners' discussions did not lie in an agreed set of intrinsically convincing general rules or principles [but] lay in a shared perception of what was specifically at stake in particular kinds of human situations (Jonsen and Toulmin, The Abuse of Casuistry, 1988) (emphasis in original).

Jonsen and Toulmin's assertion that the principles of autonomy, justice, and beneficence are too general to dictate specific moral positions is correct. But indeterminacy is also a characteristic of "general principles" in religion and atheism that Jonsen and Toulmin cited as responsible for the commissioners' varying justifications of their moral positions. In fact, Jonsen and Toulmin present no evidence of religious influence on any commissioner's position, and the two statements of support for the recommendations do not indicate conflicting (or any) religious impact. The commissioners' disagreement about supporting statements was not substantive; it was based on rhetorical rather than religious preference.

Still, Jonsen and Toulmin's casuistry supplies an element missing from a strict principlist description of ethics. Casuistry emphasizes bottom-up reasoning based on similar cases serving as examples, rather than top-down reasoning from principles, or analogical rather than deductive reasoning. These modes of reasoning are not mutually exclusive: principles may establish the parameters of precedential examples, and examples may assist interpretations of principles. Indeed, examples may congeal into principles. This synergy of principles and precedents is familiar to lawyers, who find authority both in statutes and precedential cases. In both law and ethics, consistent treatment of similar situations is required, and the question whether situations are similar or distinguishable is often debated.

From my lawyer's perspective, however, casuistry was no more successful than principlism either in generating or rationalizing the commission's recommendation on minor-increase-over-minimal-risk research with children. This research confronted the commission with the concerns customarily posed by human research: whether potential benefit to others (i.e., in nontherapeutic research) balances risk to subjects, whether parents' responsibility bars their consent to risk on behalf of their children, and whether the selection of children to participate in research is fair. To resolve these issues or, more precisely, to develop criteria for IRBs to resolve these issues in particular cases, the commission did not rely on formal reasoning. As noted, the basic principles were too general for this task. Similarly, precedential examples were nonexistent or too distinguishable to support a solution. The resolution came instead from a more common but less definable source of human value: practicality.

The commissioners determined as a practical matter that an IRB might find the potential benefit of the research justified the accompanying risk, that parents could take this favorable comparison into account in consenting to their children's participation in the research, and that selecting children with the condition under study was fair. General principles established the boundaries – potential benefit must justify potential harm, parental responsibility must be respected, and there must be a valid basis for selecting the participants – but (as Jonsen and Toulmin observed) these principles did not dictate the details of the commission's recommendations, including, crucially, the acceptance of a minor increase over minimal risk in nontherapeutic research. Further, although parents' consent to their children's participation in risky sports was noted, this was hardly a controlling example, nor was there any other example to be followed in the context of research with children. The convincing consideration was practicality: members of the commission were concerned that strictly barring children's participation in nontherapeutic research unless the research presents no more than minimal risk or is approved by a national ethical advisory board would add only a modicum of protection at the cost of gaining information needed to develop therapies for currently untreatable illnesses of children.

For dissenting commissioner Cooke, a justification based solely on utility was problematic, indicating the recommendation was "perilous" and the commission "ethically uncertain." But Cooke's criticism was undercut by his preference for a national ethical review body to review nontherapeutic research presenting more than minimal risk to children. Elevating the decision-making process to a more elaborate, public, and potentially less conflicted review would not cure an objection to the commission's utilitarian justification. Further, utility was the commission's primary but not its sole justification. The commission limited the IRBs' remit to a "minor increase" and also recommended conditions to assure meaningful parental consent and fairness in the selection of participants.

Utility is clearly a necessary, though insufficient, justification for the conduct of human experimentation. Sufficient justification requires, in addition, respect for persons, fairness, and possibly other conditions. It is virtually a truism that human experimentation, which consumes social resources and may expose individuals to harm, should return a potential benefit that outweighs the potential harm. How to measure utility (especially when the potential benefit and harm accrue to different individuals) may in some cases be a complex matter subject to disagreement, but the bare requirement of utility is hardly disputable. We may disagree and debate whether a research project is likely to produce a net benefit, but it seems beyond debate that approval of the research should be based on a determination that it is likely to produce a net benefit. Indeed, the assertion of Jonsen and Toulmin that the "locus of certitude . . . lay in a shared perception of what was specifically at stake" hints at an element of pragmatism in their analogical framework.

2.7 The Belmont Report

The commission is remembered primarily for its Belmont Report (National Commission for the Protection of Human Subjects, 1978), which presents principles and corresponding guidelines for the ethical conduct of human experimentation. These standards have played a significant role in the broad recognition of fundamental ethical requirements for human research and, by extension, the practice of medicine. Without diminishing its importance, however, the Belmont Report has not resolved many difficult issues in bioethics. Common acceptance of the Belmont principles does not necessarily lead to their uniform application in the face of many potential causes of disagreement: different interpretations of the principles, conflicts between principles, enforcement issues, and perhaps most contentious, disagreement about relevant facts and consequences. Knowledgeable persons and, indeed, the commission itself have considered the role of the Belmont Report as more to raise questions with its compendium of concerns and considerations, than to resolve issues.

The commission retreated to Belmont, a colonial manor in rural Maryland, to consider its legislative mandate to "identify the basic ethical principles which should underlie the conduct of biomedical and behavioral research involving human subjects [and] develop guidelines which should be followed in such research to assure that it is conducted in accordance with such principles" (U.S. Congress, 1974). This assignment, unlike others in the commission's mandate, did not focus on a particular population of research subjects who might be unable to consent, but on human experimentation in general. Also, the commission's task of identifying principles had no empirical component, aside from gathering scholarly nominations of the principles to be included. In contrast to the factual investigations and analysis that played an important role in developing specific regulations for research with the fetus, prisoners, and children, the Belmont process was more introspective and culminated in a general policy statement. When the Belmont Report was issued, I anticipated little public interest in this abstract document. Who knew?

Contrary to my expectation, the Belmont Report, named after its birthplace, became the commission's best known product and a foundational resource in bioethics. Its influence has been pervasive. Although never formally promulgated, the Belmont Report became the de facto federal policy for protecting human research subjects and the explicit policy of virtually every Institutional Review Board, by virtue of its inclusion in the template for institutional assurances (Office for Human Research Protections, 2008). As a result, institutions with IRBs proclaim that the Belmont principles will guide all their human subjects research. (Indeed, an alleged failure to follow the Belmont principles adopted in an institutional assurance was a cause of action in an unsuccessful law suit against the Hutchinson Cancer Research Center in Seattle.) Further, the Belmont principles have "permeated" clinical medicine (Cassell, 2000), and the spirit of Belmont in quadripartite format, as well as scraps from the commission's cutting-room floor, live on in Beauchamp and Childress's influential Principles of Biomedical Ethics (Beauchamp, 2005).

The concise Belmont Report (20 double-spaced pages in its original typewriter font) elucidates three general principles for ethical conduct of human experimentation and three corresponding guidelines for implementing the principles. This structure follows a simple schema that I suggested to Tom Beauchamp, a principal drafter of the report (Beauchamp, 2005):

Principle of	Applies to	Guidelines for
Respect for persons		Informed consent
Beneficence		Risk-Benefit Assessment
Justice		Selection of subjects

The explication of these principles and guidelines in the Belmont Report achieved the commission's goal of crispness (Jonsen, 1998) and provides a succinct introduction to ethical doctrine relevant to human research.

The Belmont Report had substantial precedent. Congress specified that in identifying the basic ethical principles for research with human subjects, the commission should consider risk-benefit assessment, selection of subjects, and informed consent. (U.S. Congress, 1974) These three concerns, which became the "applications" of the Belmont principles, had already been recognized in the Nuremberg Code (1947) and Declaration of Helsinki (1964), as well as HEW's "Yellow Book" (Institutional Guide to DHEW Policy on Protection of Human Subjects, 1971). In the Belmont Report, the commission took these documented concerns and worked backward to supply an underlying rationale and forward to elaborate the implications for ethical review.

Interestingly, the commission did not claim to have identified every relevant principle, but three of sufficient breadth and generality to assist understanding of the ethical issues in human research. Further, the commission acknowledged that its three principles would not necessarily be conclusive:

> These principles cannot always be applied so as to resolve beyond dispute particular ethical problems. The objective is to provide an analytical framework that will guide the resolution of ethical problems arising from research involving human subjects (National Commission for the Protection of Human Subjects, 1978).

Thus, the commission envisioned a didactic and methodological, rather than a dispositive, role for the three principles. An "analytical framework" is not a set of axioms or premises from which the resolution of ethical issues can always be deduced. Rather, it establishes general requirements to be met, and concerns and considerations to be taken into account in the ethical analysis of human research.

As in legal analysis, the application of the Belmont principles in particular cases may require inter alia an interpretation of broad terms, investigation of relevant facts, balancing of conflicting standards, and consideration of consequences. To take an example common to law and ethics, many discriminatory acts are clearly unfair, but some discriminatory acts – the sort that generate legal cases and ethical arguments – may raise contentious issues about fairness notwithstanding agreement on the basic prohibition of unfair discrimination. Thus, multiple opinions in a Supreme Court case may agree that unfair discrimination should be prohibited but reflect a variety of views and a split decision about such matters as the meaning of fairness, the significance of various factual situations (e.g., differential impacts of facially neutral actions), the rights of those adversely affected by anti-discrimination measures, and the likely consequences of following one or another determination of these considerations. A general principle or imperative prohibiting unfair discrimination, then, will support a wide range of results depending on many considerations involved in applying the principle. Indeed, the principle may provide only the starting point and, ultimately, a convenient, if not conclusive, rationale for an ethical or legal determination.

The commission's largely empirical approach to ethical issues and its invocation of procedural rather than substantive means to resolve some issues demonstrate the limited role of the Belmont principles. Although the commission devoted substantial effort to identifying and enunciating principles, the members of the commission generally sought empirical reasons and justification, focusing more on likely consequences than abstract grounds for its recommendations. The commission's empiricism was the fulcrum of its policy determinations, which might fit various theoretical explanations.

Today there is no lack of principles and rules to guide or regulate human experimentation. But the looming presence of these standards may imbue ethical review with an abstract quality, leading to an overemphasis on nominal compliance with rules and standards, while insufficient effort is devoted to evaluating the factual details and the impact of measures to protect human subjects. The focus on rules and principles may crowd aside a broader, pragmatic perspective that emphasizes an empirical review of the proposed research activities, an evaluation of ethical requirements in terms of their actual effects on the human subjects, and a skeptical attitude toward highly speculative risks and benefits. (Wolf, 1994) Rules and principles cannot be disregarded; they provide useful guidance. They are starting and endpoints, but the ethical review of research is an empirical, investigational task, not merely an "application" – whatever that signifies – of abstract rules and principles.

References

Advisory Committee on Human Radiation Experiments. 1996. Final report. New York: Oxford University Press.

Alexander, S. 1962. They decide who lives, who dies. *Life* 53: 102–125.

Beauchamp, T.L. 2005. The origins and evolution of the Belmont Report. In *Belmont revisited*, eds. J.F. Childress, E.M. Meslin, and H.T. Shapiro, 12–26. Washington, DC: Georgetown University Press.

Beecher, H.K. 1966. Ethics and clinical research. *New England Journal of Medicine* 274(24): 1354–1360.

Brody, J., and E. Fiske. 1971. Ethics debate set off by life science gains. *New York Times*, March 28.

Cassell, E.J. 2000. The principles of the Belmont Report revisisted. *Hastings Center Report* 30(4): 12–21.

Department of Health, Education, and Welfare. 1971. The Institutional Guide to DHEW Policy on Protection of Human Subjects.

Goodman, W. 1967. Doctors must experiment on humans but what are the patient's rights? *New York Times Magazine*, July 2.

Jonsen, A.R. 1998. *The birth of bioethics*. New York: Oxford University Press.

Jonsen, A.R., and S. Toulmin. 1988. *The abuse of casuistry*. Berkeley, CA: University of California Press.

Mahoney, M.J. 1975. The nature and extent of research involving living human fetuses. In Appendix, Report and Recommendations – Research on the Fetus, by National Commission for the Protection of Human Subjects.

Mitford, J. 1973. Experiments behind bars: Doctors, drug companies and prisoners. *Atlantic Monthly* 23, January: 64–73.

National Commission for the Protection of Human Subjects. 1975. Report and recommendations: Research on the fetus. Washington, DC: U.S. Government Printing Office.

National Commission for the Protection of Human Subjects. 1976a. Report and recommendations: Research involving prisoners. Washington, DC: U.S. Government Printing Office.

National Commission for the Protection of Human Subjects. 1976b. Research involving prisoners: Appendix to report and recommendations. Washington, DC: U.S. Government Printing Office.

National Commission for the Protection of Human Subjects. 1977. Report and recommendations: Research involving children. Washington, DC: U.S. Government Printing Office.

National Commission for the Protection of Human Subjects. 1978. The Belmont Report. Washington, DC: U.S. Government Printing Office.

Office for Human Research Protections. 2008. Terms of the Federalwide Assurance (FWA) for the Protection of Human Subjects. http://www.hhs.gov/ohrp/humansubjects/assurance/filasurt. htm.

Rothman, D.J. 1991. *Strangers at the bedside*. New York: BasicBooks.

Toulmin, S.E. 1975. Fetal experimentation: Moral issues and institutional controls. In Appendix: Research on the Fetus, by National Commission for the Protection of Human Subjects.

U.S. Congress. 1973. Congressional Record. Sept. 11. S16348.

U.S. Congress. 1974. Pub. Law 93–348.

U.S. Congress, Committee on Labor and Public Welfare, Subcommittee on Health. 1973. Quality of health care – Human experimentation, 1973. Hearings. Vols. 1–4.

Wolf, S.M. 1994. Shifting paradigms in bioethics and health law: The rise of a new pragmatism. *American Journal of Law & Medicine* 20: 395–415.

Chapter 3
Genesis of a Totalizing Ideology: Bioethics' Inner Hippie

Griffin Trotter

3.1 Introduction

As 1974–1975 newspaper editor for Rogers High School, south of Puyallup, Washington, I helped fashion the popular "student of the month" feature. Each month we chose some special student from amongst our repertoire of rural, working class offspring. A reporter was then dispatched to administer an array of sophisticated questions, including: "What is your philosophy of life?" The uniformity of response was striking. Virtually everyone articulated the same probing personal creed: "Do your own thing, as long as it doesn't hurt anyone else." I have come to believe that this moral mantra was hardly unique to our local sample of emerging humanity. To the contrary, it communicates a fundamental conviction we shared with most others of our generation – something old enough to be entrenched by the mid-seventies, yet sufficiently newfangled to resonate among the restless, unformed youth of that time.

Fast forward to the present day. Now I am an academic bioethicist, not involved in "bioethicist of the month" features, but still stricken by the uniformity of moral thought among my peers – and their tendency to believe that they have arrived at some kind of profoundly wise collective consensus that portends the transformation of others who are not so enlightened. Like its predecessors at Rogers High School, this current crop of peers is awash with confidence but uneven in its understanding of history, economics, and the general dynamics of social transformation. Further, it exhibits a youthful obliviousness about self-contradiction. Unlike its predecessor, it possesses a political ideology – a quality that makes it more practically potent, but also potentially more dangerous.

G. Trotter (✉)
Albert Gnaegi Center for Health Care Ethics, Salus Center, Saint Louis University,
St. Louis, MO 63104, USA
e-mail: trotterc@slu.edu

H.T. Engelhardt, Jr. (ed.), *Bioethics Critically Reconsidered*,
Philosophy and Medicine 100, DOI 10.1007/978-94-007-2244-6_3,
© Springer Science+Business Media B.V. 2012

In this essay I explore the ideology of contemporary bioethics and its basis in values that crescendoed in the 1960s, when most of today's most influential bioethicists grew up or cut their academic teeth. These are values that informed my high school classmates and continue to inform most of the generation of adults to which we belong – especially those who reside in the academy. These are not by any means the only values in the mix, but I will claim that they are critically important. I call them "hippie values," but they are not inherited solely or even primarily from the hippies – and they certainly did not originate with them. They came to great prominence during the sixties, and derived some of their public notoriety through the activities of hippies in the middle of that period. But I was no hippie and neither were most of my colleagues in bioethics. Many of us didn't even like hippies. To the contrary, our "hippie values" were acquired through a number of pathways, including the "New Left" politics of student groups such as Students for a Democratic Society, mainstream politicians such as Eugene McCarthy and George McGovern and, most influentially, from potent media of cultural transmission such as popular entertainment (especially rock music), youth chic, and just about anything else that helped us establish our own identity vis-à-vis the perceived moral-cultural rigidity and hypocrisy of our parents' generation.

Many scholars have already attempted to isolate the historical content, sources and embodiments of such "hippie values," and I will not duplicate, re-create or summarize their findings here. When I allude to particular historical influences, I will not confine myself to the hippies, nor will I restrict myself to views that were consciously approved by that small and mostly apolitical group. I will look broadly, and focus on influences from the cultural revolution of the 1960s that seem to have penetrated most deeply into the psyche of contemporary bioethics (most definitely not an apolitical field). At the same time, I contend that there is something special about the hippies that makes them a fitting emblem for contemporary bioethics – and not merely because of the strong ideological links I will soon discuss.[1] Like today's bioethicists, the hippies were predominantly white, and raised with unprecedented personal freedom and luxury. This contrasts with many of the prime movers in the civil rights movement, who stepped up from brutal oppression. We bioethicists may not be as self-inflated as prepsters Abbie Hoffman and Timothy Leary, or as pampered as New Left icons Susan Sontag and Jane Fonda, but most of us came from somewhat similar backgrounds – light years removed from the experience of a Martin Luther King, Jr.

My thesis is that the hippie legacy, interpreted broadly, lurks in bioethics' collective unconscious and manifests in dogmas, contradictions and totalizing aspirations that currently plague the field. Three roughly sequential but overlapping denominations of this legacy – one involving an escape from normalcy, one featuring the rhetoric of love and the third featuring the politics of rage – will be examined and linked to movements of thought in bioethics. I will argue that these denominations are not only mutually incongruent, but that each also exhibits internal tensions, and that these problems persist in the current manifestations.

3 Genesis of a Totalizing Ideology: Bioethics' Inner Hippie 51

3.2 The Escape from Normalcy: "Do Your Own Thing"

The cultural revolution of the 1960s resulted to some degree from an identity crisis among the youth of that decade. Psychologically speaking, there were features of the typical "identity deficit" crisis of adolescence, precipitated by: (1) repudiation of parental values, (2) cognitive advances allowing youths to exert greater control over their identities,[2] and (3) the pressure to choose between various potential identities (Baumeister et al., 1985; Levi et al., 1972). These reactants were catalyzed by the Vietnam War, which posed a particular threat among those susceptible to conscription. For many young persons, the politics of war recapitulated an authoritarian regime at home – where they felt constantly held at bay by rigid, controlling parents who expected them to parlay their unprecedented educational opportunities into dubious careers.[3] In the wake of Joseph McCarthy and the red scare, the vaunted nationalistic ideals of the World War II generation came under suspicion. Likewise with parents' fidelity to the "American Dream," which was viewed by some youth as a form of slavery to corporate America.

To successfully navigate an identity deficit crisis, one must differentiate oneself from others, especially from those who served earlier as models. At the same time, one must establish new forms of social continuity, yielding identity components sufficient to fill the deficit. The escape from normalcy became a prominent medium for the first of these tasks. Whatever else you can say of them, the rebellious sixties youth were determined not to become their parents.

Philosophically speaking, what evolved in the early 1960s was a new conception of individuality – focusing on the escape from social normalcy and the exploration of one's inner self. To achieve such authentic interiority, one needed to shrug off the received tradition, deflect indoctrination, and free oneself of social expectations. The new individuality proposed a version of the self that lacked such encumbrances. The problem, of course, is that the removal of encumbrances left little basis for a stable identity. With a newly expansive menu of identity-forming options, identities were tried on and swapped out with unprecedented speed.

The motto became: "Do your own thing." And most of the things that young people did involved some kind of self-exploration through sensual indulgence. Such activities quickly coalesced into a cultural movement – what we will cover in the next section on the rhetoric of love – with its own set of values, expectations and meta-criteria for identity formation. Before moving on, however, it is useful to examine some features of contemporary bioethics and value theory that reflect this phase of intensive individualism.

One immediately evident parallel is the emphasis in bioethics' early years on autonomy as radical choice – over and against physician paternalism. The autonomy movement has been enshrined in bioethics' canonical works and remains prominent despite increasing misgivings about its excesses. As Beauchamp and Childress note: "Virtually all theories of autonomy agree that two conditions are essential for autonomy: (1) *liberty* (independence from controlling influences) and (2) *agency* (capacity for intentional action)" (2001, p. 58). Where these theories run into trouble – especially from the standpoint of the escape from normalcy – is in specifying just what should count as an untoward "controlling influence," and how much continuity of character is required for authentic agency. Is it harmful when children or young adults defer to parents' values? Or when a patient decides on the basis of the values of a particular social group? Extricated from such values, on what basis is the patient to establish enough continuity of character to choose authentically (rather than on the basis of a fleeting impulse or intuition)?

These questions reflect Josiah Royce's conception of the paradox of self-realization (Trotter, 1997, pp. 57–58). He writes:

> Here, then, is the paradox, I and only I, whenever I come to my own, can morally justify to myself my own plan of life. No outer authority can ever give me the true reason for my duty. Yet I, left to myself, can never find a plan of life. I have no inborn ideal naturally present within myself. By nature I simply go on crying out in a sort of chaotic self-will, according as the momentary play of desire determines (Royce, 1995, p. 16).

This paradox was accentuated in the sixties escape from normalcy, because the fleeting nature of sensual indulgence, and the non-fixity of values it begets, are in tension with the psychological need to establish a stable personal identity. That tension remains today as a liability in bioethics' focus on individual autonomy.

It also vitiates health policy theory. Currently there is interest among many bioethicists in applying John Rawls's conception of justice as fairness to problems of healthcare distribution. Rawls asked us, in 1971, to imagine deliberators in an "original position" where they know nothing of their place in society, their temporal circumstances, their fortune in the distribution of natural assets and abilities, their life plans or their conceptions of the good (Rawls, 1971, p. 12). If we do, he thought, we will see that deliberators in this hypothetical position would choose a particular, political conception of justice as fairness as the basis for a scheme of cooperation. Interestingly, Rawls adopts Royce's conception of a person – as "a human life lived according to a plan" (Rawls, 1971, p. 408; Royce, 1995, p. 79) – which implies that the entities in the original position are not persons. Somehow, Rawls thinks that their native rationality, combined with knowledge of certain general facts (not including their historical time of living), would be enough to beget unanimity about a political conception of justice, and that this common conception would, in turn, determine a higher-order personal identity that serves as a meta-criterion for other identity components, including the selection of a life plan (Rawls, 1971, pp. 136–150, 563).

The resulting conception of justice (we will not examine its details), and its spin-offs in bioethics, is plagued by the problem of un-encumbered individualism. As several critics have noted, Rawls's proto-person in the original position is largely devoid of identity-forming components, which are the true bearers – not only psychologically, but arguably also epistemologically – of stable values. Thus deprived, Rawls's deliberators seem to have insufficient rational basis for preferring one plausible version of justice over another in the original position.[4] Rawls would counter that some identity-forming components (conferring part of our sense of continuity) derive from our common rational nature, and hence are admissible behind the veil of ignorance. The problem is that this common rational nature, in whatever degree it actually exists, is hard to fix and seems to under-determine a political conception of justice. Furthermore, Rawls's theory requires that we should and rationally would be willing to assent to a theory of justice while subjected to ignorance not only about our own ultimate ends, but also about truths that presumably might structure such ends – including the truth about good and evil, the ultimate meaning of life, the will of God, and so forth. And he requires us to use this conception of justice as a meta-criterion for personal identity, effectively closing us off from certain plausible inquiries about the aforementioned, ultimate truths. In other words, Rawls postulates a political ontology of personhood in which commitment to a conception of justice supercedes and limits the pursuit of other moral truths and the formulation of a life plan. Rawlsian personhood, then, conceived not in its fullness as Royce's person-with-a-life-plan, but in its more fundamental, stripped-down form as a disassociated seat of choice, seems to carry quite a bit of metaphysical baggage.

Thomas Nagel complains that the original position manifests "a strong individualistic bias ..." and "seems to presuppose not just a neutral theory of the good, but a liberal, individualistic conception according to which the best that can be wished for someone is the unimpeded pursuit of his own path, provided it does not interfere with the rights of others" (Nagel, 1975, pp. 9–10). On Nagel's account, then, Rawls assumes we all share the personal philosophy articulated by students at Rogers High School at about the same time Rawls wrote his book. This may overstate the case a bit (since Rawls allows for an array of world views and community associations that would repudiate such an ethos), but Nagel is unquestionably right to identify and criticize the primacy of individual choice in Rawls. We should also contest the version of the self that provides a metaphysical locus for Rawlsian choice.

These problems are exacerbated when Rawlsian justice is injected into healthcare. Norman Daniels, for instance, proposes to include healthcare institutions and practices among the basic institutions regulated by Rawls's notion of justice as fairness (Daniels, 1985, p. 45). Since our opportunities hinge to some degree on our good health, Daniels thinks that equality of opportunity presumes some kind of guaranteed access to healthcare. Critics might charge that health is a complicated concept, much more laden with contingent moral beliefs than Rawls's "basic goods," and hence cannot be politically specified apart from a thicker-than-allowable, general conception of the good. To this, Daniels replies that "persons are not defined by a particular set of interests but are free to revise their life plans. Consequently, they have a fundamental interest in maintaining conditions under which they can revise their life plans as time goes on" (Daniels, 1985, p. 47).[5] To allow for a normal range of possible life plans, one must presume a normal range of functional capacities. And this begets Daniels's conception of health as normal species functioning

Now it is contestable to presume, as Rawls does, that entities in the original position can maintain a capacity to distinguish between right and wrong while being so clueless about good and evil and related matters that they will have an overriding interest in preserving the opportunity for as large as possible an inventory of potential life plans. But Daniels, invoking Rawls, wants to add the infinitely more contestable claim that fully embedded actual adults must be presumed to bear the same interest, and this interest is a more fundamental aspect of their personhood than their commitment to any particular conception of the good or life plan (notwithstanding that Rawls, contra Daniels, defines the embedded person by his/her commitment to a life plan). But do Dominican priests really have a more fundamental interest in preserving the opportunity to become abortionists than they do in serving God? Is preserving the opportunity to play hockey really more important to a young artist than such life-plan contingencies as learning to draw the human figure? Is maximizing life options even psychologically healthy?[6] The answers are affirmative only if living rationally and healthily are more about preserving choice than about realizing particular ends. Or, to put it differently, life is fundamentally a matter of "doing your own thing" for as long as, but no longer than, the spirit so moves you. What spirit? To that question we now turn.

3.3 The Rhetoric of Love: "Make Love, not War"

A release from social constraints is never enough to resolve an identity crisis. Though it helps with the task of differentiation and may open an array of life options, it does not alone provide the social continuity requisite for a stable personal identity. In other words, it leaves an identity deficit. That is why the escape from normalcy was at best only a beginning for 1960s youth. And, given the plausibility of certain linkages between natural psychology and the ontology of persons, it is also one means (but not the only[7]) of explaining why a social philosophy that enshrines unencumbered choice can never succeed as a foundation for bioethics or political theory.[8]

Sensual experiences are attractive to experimentally inclined youth, but, as noted in the last section, do not of themselves provide a sufficient framework for identity formation. To establish some sense of social meaning or purpose in life (I presume that such is a fundamental psychological need), the sensualist must either: (1) create or appropriate some device that imbues sensual experiences with deeper meaning, or (2) augment sensuality with some other source of social continuity. The rhetoric of love moves primarily along the first of these paths, though its evolving anti-war sentiments lay groundwork for the second path (covered in the next section).

In 1966, Norman O. Brown, regarded by some (Matusow, 1984, pp. 277–280) as the intellectual prophet to the hippies, wrote:

> The solution to the problem of identity is, get lost ... Dionysus, the mad god, breaks down the boundaries; releases the prisoners; abolishes repression; and abolishes the *principium individuationis*, substituting for it the unity of man and the unity of man with nature (Brown, 1966, p. 161).

This simple formula expresses the essence of the hippie movement as it is embodied in the rhetoric of love. Contra Freud, Brown holds that the ego is fundamentally a unity, and that this unity is Dionysian in nature; it is Eros.[9] The ego that wants to find itself must first lose its socially constructed identity through sexual indulgence. Subsequently it will achieve an ecstatic union of body and soul with other Dionysian egos.

3 Genesis of a Totalizing Ideology: Bioethics' Inner Hippie

The problem of identity formation, then, is solved on this account through the establishment of a perfect, sensual unity that excludes individual identity. By doing its own thing, the self eventually erases itself, and enters an enlightened state of perpetual unity, harmony and erotic love. The process of differentiation thus begets continuity – leaving differentiation behind as a relic of the old establishment. In the process, the tension between individualism and sociality is resolved mystically through sex and (in subsequent accounts) its special sacrament: hallucinogenic drugs. And, of course, there are hymns – of the rock "n" roll variety.

This creed was brought to the masses by media personalities such as Timothy Leary.[10] His popular slogan, "Tune in, turn on, drop out," traces the social trajectory of the proposed psychological journey. He also succinctly describes the destination and mode of transportation: "Blow the mind [through hallucinogens] and you are left with God and life – and life is sex" (Matusow, 1984, p. 290). As for rock "n" roll, Leary claimed after hearing the Sergeant Pepper album that the Beatles were "evolutionary agents sent by God" (Matusow, 1984, p. 296).

Needless to say, in the long run this religion of endless, uninhibited sex and drugs did not prove to be as fulfilling as converts initially hoped. There were downers and overdoses. There was sexually transmitted disease, hunger and various persistent physical discomforts. Ultimate harmony notwithstanding, Haight-Ashbury wallowed in violence and property crime. And mystical clarity, if it was achieved at all, tended to be ephemeral. Though its proponents fancied themselves as spiritual siblings to Buddhists, Hindus and native Americans, the sixties religion of erotic love lacked the social infrastructure and cultural traditions that buttress these other, more durable forms of spirituality. Perhaps more to the point, its adherents lacked the self discipline required to adhere to ascetic and quasi-ascetic regimes that are the true source of spirituality in such religions (contra the hippies' claims about sensual self-indulgence). They were, borrowing a word from Roger Kimball, stuck in a state of perpetual "babydom."[11] Thus, writes Matusow,

> when [Allen] Ginsburg and Richard Alpert met Hopi [Native American] leaders in Santa Fe to propose a Be-In in the Grand Canyon, the tribal spokesman brushed them off, saying according to the *Berkeley Barb*, 'No, because you mean well but you are foolish ... You are a tribe of strangers to yourselves' (Matusow, 1984, p. 298).

The hippie synthesis of radical individualism and radical social unity fails because the individualism of self-indulgence only multiplies the selfishness and divisiveness that give it birth. Apart from moral commitments that run deeper than slogans about love and peace, there is little basis for the tremendous convergence of moral sentiment that the hippie utopia requires. There is no universal ethos that magically presents and sustains itself in the aftermath of ecstatic sex or mind-blowing drug adventures. In their state of cultivated infantilism, hippies were able for a time to sustain their belief in a mystical harmony induced through sexual union. But their personal demons, far from being vanquished, built up. And, in the midst of a civil rights movement with origins elsewhere, some of the more thoughtful dropouts seem to have realized that their bondage to the establishment was trivial in comparison to massively oppressed groups such as African Americans. Some demons, it became apparent, are extricated only through active resistance – and sometimes only through force of arms. Such insights contributed to the next, political denomination of hippie values – values that take us beyond the brief lifespan of the hippie movement itself.

To my knowledge, no contemporary bioethicist subscribes to the sensual religion of the hippies. Its remnants can be found in contemporary New Age spirituality, which is much sanitized from the standpoint of pan-sexuality, but nearly as inane in its basic tenets as the prior Dionysian religion. Bioethics is aloof from New Age religion as well.

Where the sixties rhetoric of love penetrates most deeply in bioethics is, first of all, in bioethicists' insistence, vis-à-vis an overwhelming, contradictory pluralism of incompatible moral communities, that large nations such as the United States harbor sufficiently convergent moral values to be regarded as moral communities (bearing, for instance, common moral standards of justice, patient autonomy, and so forth). Some bioethicists even hold that humanity itself is morally unified to a degree sufficient to establish a global bioethics (exemplified, for instance, in universal declarations of bioethical values by organizations such as UNESCO). Extant rebuttals to such universalism are numerous, detailed, compellingly argued – and roundly ignored by the mainstream bioethicist, who rivals hippies in the ability to suspend disbelief in the morally fabulous. In analogy to the hippies, mainstream bioethicists believe – apparently on faith alone – that their emphasis on individual rights and patient autonomy will soon yield a universal moral consensus that binds disparate regions and moral communities.

A second important bioethical remnant of the rhetoric of love is the urgency to normalize, or "de-stigmatize," deviant or irresponsible expressions of sexuality. "Promiscuity" and "prostitution" are removed from the bioethical dictionary. Instead we have "partner changing" and vulnerable populations of "sex workers" (a clinically ambiguous category that presumably would include not only men and women of the night, but also less-STD-prone occupations such as pornographic movie director, dildo salesperson and sex therapist). Without question, some of the sentiment against stigmatization is well founded, and has produced benefits for individuals susceptible to serious illnesses such as HIV infection. But bioethicists' typical categorical dismissal of virtually any call to sexual responsibility, on the basis that it manifests oppressive moralism, is incongruent with bioethics' moralistic approach to tobacco use, firearms ownership, entrepreneurism and, well, just about any potentially unhealthy practice that was condoned by the old establishment (Trotter, 2003b).

For instance, Arthur Caplan (2005) writes that there "may be a sillier strategy for dealing with sex among teens than promoting the choice of "abstinence-only-until-marriage," but I am not quite sure what it is." He derides abstinence not only because it won't work (he seems correct about the futility of school-based programs currently in favor with the Bush administration), but because it is unethical. He continues:

> The message that sex must wait until marriage is not the right message to send to a young person. The people sending the message almost never lived up to it in their own lives and nothing turns off a kid like hypocrisy. Furthermore, most kids themselves just don't believe it.

3 Genesis of a Totalizing Ideology: Bioethics' Inner Hippie 57

Features of Caplan's argument bear noting. First, he declares that one of the central moral teachings of Christianity, Judaism, Islam, and other traditional religious groups is silly. Second, he bases his argument on an implicit claim that anyone who fails to live up to a moral ideal is a hypocrite if they continue to profess that ideal. This would not bode well for most of the other ideals of traditional morality, including altruism, truthfulness, and respect for parents – ideals that for many religious persons, in their alleged mist of confusion, seem to remain worthy even if they are difficult. It also implies that alcoholics (recovered or otherwise) who advise their children against drinking, and smokers who advise their children against smoking, are ethically misguided. Third, he implies that the sexual morals of American youth are relatively fixed, perhaps even by nature itself. Yet data from the Population Reference Bureau (2008) indicate that the number of unmarried persons 15–19 years of age who have had sex varies greatly between countries, from over 60% to less than 1% (data for females aged 15–19), and that the United States has much higher than average rates of teenage sexual activity (though presumably they would move toward the middle if more European countries were included in the analysis).

Instead of scrutinizing the ubiquitous sexual images in popular entertainment, and linking them to the obvious fact that classroom-based "abstinence-only" sex education fails because no one looks to their high school health teacher for guidance on how to be cool, mainstream bioethicists such as Caplan instead deride the very notion of counseling youngsters to abstain, chanting the public health mantras that high rates of adolescent sexual activity are "simply a fact of life" or "an inevitable response to material and social deprivation" that we need to address primarily through the prevention of harmful effects.[12] Apart from the fact that both of these claims are clearly false, it is interesting to speculate on what would have happened if bioethicists and public health workers had resigned so quickly when initial efforts at classroom-based smoking abstinence education failed.

In point of fact, the academic mainstream – including bioethics – regards satisfying sexual intercourse as a fundamental component of psychological health, as crucial for unmarried young adults (or even older adolescents) as it is for adults in committed relationships (Kelly and Schwartz, 2007). This belief is neither based on, nor refuted by, available evidence. Scientific research does not confirm the claim that sexually active singles are better adjusted, healthier, or more successful than those who remain celibate. To the contrary, the tenacious rhetorical linkage between sexual intercourse and good health is a remnant of the rhetoric of love, with its doctrine that sexual expression is a more central aspect of human flourishing than, say, learning to delay gratification or expressing moral virtue. When bad things happen because of sexual indulgences, they typically are regarded as inevitable results of authentic human living.

Another, classic example is the argument by Judith Jarvis Thomson, canonized in bioethics anthologies, that the moral permissibility of abortion can be established without resolving the question of the personhood of the fetus (Thomson, 1971). Thomson claims that the condition of a woman who learns she is pregnant is analogous to that of a person who awakens to find she is connected by a lifeline to a concert violinist. If the lifeline is disconnected, the violinist dies. Though it may be laudable to keep the lifeline in place, Thomson argues, this option is supererogatory. There is no fundamental moral obligation to do so, because a person's body is her own property and cannot be appropriated by others against her will. Likewise, she concludes, even if the fetus is a person, there is no obligation for a pregnant woman to carry the pregnancy to term. The problem – amazingly not immediately evident to all readers – is that her violinist-lifeline narrative does not include an earlier interlude where the woman willingly participates in activities that are known to create lifeline connections with concert violinists (the woman in Thomson's narrative has been kidnapped by supporters of the violinist, and hooked up without her knowledge). Yet presumably most women who become pregnant have willingly engaged in sexual intercourse – known even among sub-doctoral individuals to be a cause of pregnancy. Except for cases of rape and extreme sexual manipulation, Thomson's analogy would seem to break down entirely. It does not break down, however, in the minds of individuals (presumably including Thomson and those who find merit in her essay) who think that sexual intercourse is so inextricably merged with the automatic physiology of human organisms that it is not much more susceptible to voluntary regulation than a fast heartbeat.[13] If tachycardia should not be regarded as symptom of moral decline or irresponsibility, then on their view, neither should unwanted pregnancy.

Still, even if bioethics has assimilated a watered-down dogmatics of sexual expression into its already inconsistent, broader worldview, it is clear that most bioethicists reject the paradigm of sexuality as self-realization and are at odds with the non-political orientation of those who initially championed the rhetoric of love. Like erstwhile hippies who joined the civil rights movement and adopted social activism as their modus operandi, mainstream bioethicists are political creatures. We now consider this political turn.

3.4 The Politics of Rage: "Stick It to the Man"

Liberal political columnist and Middle East specialist Thomas Friedman recently mused that "Some things are true even if George Bush believes them" (Friedman, 2003). That Friedman, who is no admirer of Bush, was inclined to articulate this rather straightforward hypothesis is indicative of the ferocity and unreflectiveness with which the contemporary left has reacted to the rise of neoconservatism. Such liberal anger is especially concentrated in the academy, where few opportunities to belittle, bemoan or berate the current President are passed up. "Emotional thermometer" studies indicate that significant numbers of left liberals consider their negative feelings about Bush to be as intense as hatred can be, surpassing their dislike for Saddam Hussein (Brooks, 2008) – though presumably they might be able to achieve even higher levels of disgust if Bush emulated Hussein and dumped nerve agents on Blue states, or started hauling left liberals out in the night for a rendezvous with dismemberment, murder and mass burial.

3 Genesis of a Totalizing Ideology: Bioethics' Inner Hippie

This orientation to some degree recapitulates that of the late-sixties New Left (which, unlike the hippies, actually enlisted a generous number of individuals who later became academic bioethicists, political scientists, economists, and such). For the New Leftists, rage was a tool of protest, and perhaps the most important weapon in an emerging political struggle against oppression, discrimination, and the Vietnam War. Hence, they found it difficult to tolerate even meager gestures of approval for the United States. Former New Left activist Todd Gitlin recalls:

> Dave Dellinger told the Vietnamese at one point, 'You are Vietnamese and you love Vietnam. You must remember that we are Americans and we love America too, even though we oppose our government's politics with all our strength.' At which one American groaned and others looked embarrassed. Little by little, alienation from American life – contempt, even, for the conventions of flag, home, religion, suburbs, shopping, plain homely Norman Rockwell order – had become a rock-bottom prerequisite for membership in the movement core (Gitlin, 1993, p. 271).

Like the old New Left, today's anti-neoconservatives tend to be more focused and more passionate about what they oppose than about what they support. Both movements are strongly anti-war, yet sanguine about threats of force against out-groups who are unlikely to fight back (libertarians, business owners, smokers, etc.). They are inclined to enforce a redistributionist global ideology but opposed to American imperialism – maintaining friendships, or at least friendly openness, to mass murderers who profess the correct ideology (Mao Tse-tung, Hugo Chavez) or at least exhibit the virtue of being angry at the right people (Yasser Arafat). Misleading propaganda is openly acceptable among many members of both groups (though this is probably true of everyone, left or right, who prioritizes advocacy over inquiry). And they approach almost every social and political issue with a rigid, simpleminded hermeneutic of suspicion, neatly dividing the world into oppressors (who have free wills and therefore should be reviled for their frequent misdeeds), victims (who are short on free will, hence bearing little responsibility for misdeeds) and champions (who serve victims by protecting them and making decisions for them, which upholds victims' dignity).

With the transition to political activism, the sixties cultural revolutionaries were able to achieve a much more stable framework for identity formation. Indeed, the paradigm has survived, with a few alterations, into the present day. In this politics of rage, the linkage to sex, drugs and rock "n" roll is transformed. Sex, as noted above, is converted from a medium of universal consciousness to a prerequisite for psychological health. The love affair with illicit drugs is tempered, as these also are divested of their sacramental qualities – though, interestingly, even today left liberals tend to favor the legalization of marijuana while rejecting even relatively safe forms of tobacco use.[14] Rock "n" roll remains central, but adds a strong component of political dissent to its repertoire.

With the waning tendency to view sexual exploration and hallucinogens as spiritual media, the New Left needed either to (1) abandon its utopianism, (2) find some other medium for reconciling the tension between individualism and collective unity, or (3) discover a new kind of utopia that tempers individualism (I presume that all utopias are strong on collective unity). It adopted a loose framework that combined the second and third options. The soft sciences, especially sociology (including "socially responsive" but analytically challenged versions of economics such as Marxism), became the medium of choice for navigating the divide between individualism and collectivity. In the process, individualism was tempered, with the emergence of a new socially and politically embedded individual. This new version of the individual is one who can flourish, and express authentic individuality, only in the context of the proper, nurturing socio-political environment – an environment that celebrates sexual freedom, lifestyle pluralism, harmony with nature, transfers of wealth, and other features of the leftist good life. These qualifications come at the expense of competing values characterizing older forms of individualism – such as economic freedom, traditional religious devotion, appropriation of natural resources, and personal responsibility. As I have argued elsewhere (Trotter, 2002), this framework antagonizes the concept of toleration (which recognizes the failure of reason to resolve deep moral pluralism and requires moral strangers not to interfere with ways of life that they abhor), adhering instead to tolerance (which rejects deep moral pluralism and adopts a collective ethos featuring a superficial pluralism of lifestyles in which all persons aspire to moral friendship and are willing to support others' lifestyles). Of principle importance, then, to this emerging political ideology is the eradication or diminution of traditional morality.

The new framework reflects the image of the rebel.[15] A rebel, in the sense I intend here, is one whose personal identity hinges on an image of resistance to some object of evil or oppression. It is a potent personality ideal, that exists to some degree in most persons. It is vulnerable, however, to diminution or collapse when the object of rebellion is diminished or vanquished. In order to maintain a strong sense of personal identity and mission, the rebel must constantly discern that the evil remains strong and the fight is not yet won. At the same time, the rebel must maintain a sense of hopefulness and progress.

This is at best a difficult equilibrium, such that the historical demise of the New Left seems about equally a result of its largely successful campaigns against the Vietnam war, racism and sexism, and its tangible failures in the war on poverty and the fight against capitalism. Despite the disappearance of the New Left, the basic framework for left liberal rebellion has been nurtured and preserved, in no small part due to the intricately developed rhetorical strategy of linking its successes and failures in a narrative that reinforces the basic ideology – e.g., in claiming, despite mountains of countervailing evidence, that persistent racism, sexism and free enterprise are the root causes of persistent poverty. If the New Left is no longer a political reality, its basic ideology lives on in the academy and in the politics of the Democratic party. I will briefly examine two features of this rebel ideology that are strong in bioethics.

First is its tendency to replace reasoned argument with insistent assertion, appeals to intuition, or even intentional distortion. Caplan's aforementioned assertions about sexual activity are one example. Though bioethics tends to view itself as a bright light of reason, it has no compelling arguments for its basic dogmas (e.g., its conceptions of justice, positive rights and health, and its assumption that moral pluralism should be reduced to lifestyle pluralism and encompassed in a universal ethics).

3 Genesis of a Totalizing Ideology: Bioethics' Inner Hippie

Since these basic dogmas are the primary issues at stake between bioethics and its critics, bioethicists frequently lapse into appeals to intuition (which are effective primarily among moral friends) and the rhetoric of advocacy and protest – which hinges on the loudness and the marketability of one's assertions rather than the quality of one's logic. Science is appropriated rather than practiced. Hence, there is little effort to present data in an unbiased manner. For instance, essays and lectures on the ethical imperative for universal healthcare predictably pass over information about the economic dynamics of non-insurance, and attribute comparatively lower life spans and higher infant mortality in the U.S. (versus other industrialized nations) to lack of access to healthcare – despite the fact that access to healthcare is only weakly linked to these outcomes. There is even an attempt to describe economic inequality across society as an independent risk factor for poor health.[16]

The second feature is closely related. It is bioethics' tendency to employ the language of moral revulsion. It would be interesting to count the number of times that bioethicists have tried to establish the high ground by casting moralistic aspersions on the United States for its failure to achieve what they consider to be an equitable distribution of healthcare. The lack of universal access has been described as "embarrassing," "shameful," "reprehensible" and "deplorable" – terms that are rarely if ever used in association with policies that arguably are more health-destructive, such as a welfare system that penalizes marriage, or left-wing ideas that increase the non-insurance rate by making health insurance more expensive (e.g., laws forcing insurers to cover treatments such as infertility services, chiropractic, acupuncture, or alcohol and drug abuse rehabilitation – even in persons with good reason to believe they will never want them).[17]

Moral disapprobation is often extended even to general economic policies, such as the minimum wage. For instance, Erich Loewy charges that "decency" is lacking in the United States because it is a society "in which children are allowed to go hungry (as are 25% of children and 33% of black children in the United States), which permits a minimum wage that still puts full-time workers below the poverty line," and so forth (Loewy, 2001). His evidence and sources are omitted, but, contrary to Loewy's claims: (1) 98% of American households report having enough of the food they want (U.S. Census Bureau, 2005, p. 4)[18]; (2) though it is possible to inexpensively achieve a balanced, nutritiously adequate diet (for instance, by focusing on produce, fish and legumes), over-consumption of more expensive but less healthy foods has made obesity prevalent among the U.S. poor, (3) the poverty rate is believed by many economists to be artificially inflated and to include many households with significant wealth,[19] (4) most economists do not think that significant increases in the minimum wage are an effective policy for enhancing the well being of poor persons (Whaples, 2006; Fuchs et al., 1998), many holding that it increases unemployment, increases prices, causes better qualified individuals to leave school and displace lower qualified persons in minimum wage jobs, increases crime, increases illicit drug use, increases welfare dependency, and primarily benefits individuals in households who are well above the poverty level (Bartlett, 2000; Neumark and Wascher, 2005; Burkhauser and Finegan, 1989),[20] and (5) very few minimum wage workers remain at the minimum wage throughout a career (Carrington and Fallick, 2001). Of course, Loewy might be able to defend his remarks against these claims. The point is that if he wants to discuss his claims in some semblance of a balanced way, he should examine the counter-claims. A consensus among economists does not make something a fact – but bioethicists (like Loewy) who regard consensus in soft fields like bioethics to carry immense authority for public policy should be at pains to take account of consensus in more data-driven and analytically rigorous fields such as economics.

Balance, however, is hardly a virtue of the politics of rage. For those who define themselves in terms of the rebel ideal, the exaggeration of evil and the demonization of the opponent are necessities, not only for progress, but also for the preservation of the self and the survival of the movement. Victory has to be total – with utopia achieved – or it will not be accepted as victory at all.

Despite their inflammatory approach to healthcare reform, most contemporary American bioethicists have backed off considerably from the anti-Americanism of the 1960s. They are funded to a large extent by the federal government, dependent on the medical establishment for entry, and generally espouse a version of collectivism that would seem insipid to Marxists (in many respects like what Marx called "bourgeois socialism"). Their utopia is less extremely anti-establishment, and they are more interested in achieving it through coercive politics than through outright revolution. What remains is a basic rebel ideology that seeks political power and characterizes its opponents in angry, moralizing terms – always keen to point out that the enemy is as insidious, dangerous, wrong and strong as ever.

3.5 Conclusion

Like other totalizing social movements, mainstream bioethics is animated by a utopian social vision and seeks to establish its vision universally, by police power if necessary. Though it evolves from many sources, including the early 20th Century liberalism of Woodrow Wilson and its associated prophets (especially John Dewey), it is in large part a manifestation of the counter-cultural movement of the 1960s (which itself was influenced by prior movements). Presumably this linkage is related to the fact that most prominent bioethicists grew up in, or received their secondary education in, the 1960s.

Several denominations of the "hippie values" that emerged in the 60s have been summarized here, with a discussion of some of the tensions within and between them. The collectivist ideology attributed to mainstream bioethics is by no means the only contemporary remnant of these values. Some who were most taken with the original, individualistic manifestation have gone on, for instance, to be libertarians. And this group might be further divided into those who emphasize civil liberties and those who emphasize economic liberties. Insofar as the generation has failed to effectively resolve immense tensions between individualism and collectivity, between secularity and spirituality, and between advocacy and inquiry, such

3 Genesis of a Totalizing Ideology: Bioethics' Inner Hippie

divisions are inevitable. I have offered no reasons for concluding that the post-hippie ideology of mainstream bioethics is any less coherent or any more deluded than other extant manifestations. Nor have I claimed that "hippie values" are less coherent or cogent than their social counterparts in the establishment that preceded them. Probably they are not.

My project in this essay has been essentially negative and, in that effort, one-sided. Yet there are many positive developments from the countercultural revolution of the sixties that should be retained or even celebrated. Hippies had good reason to question the corporate culture, and were frequently insightful in their perceptions of sexual hypocrisy or gender discrimination. And if sexual indulgence and drug use were spiritual dead ends, the rock music they inspired is occasionally laced with genius.

At the same time, I believe there is an important lesson here. If we take inquiry seriously, and regard academics as more fundamentally a matter of inquiry than of advocacy for particular dogmas, then we should recognize and question our cultural inheritance from the sixties. Though rebellion persists as both a rhetorical strategy and a personality ideal, the sixties trajectory for rebellion is now so ingrained – in the academy at least – that it is not really rebellion, but rather part of the establishment. We should continue to oppose the institutions we find wrongheaded or unjust. But balanced resistance, subservient to a higher loyalty to the pursuit of knowledge and truth, and to humility about our limitations in achieving such lofty goals, is perhaps a better strategy than adopting rebellion as our primary hermeneutic and medium of self-understanding. Our rebellions might become more tentative, but they will be unleashed on a far greater array of objects – including the rigidity found in the ideology of contemporary bioethics.

Notes

1. Allen J. Matusow, perhaps the most important historian of the sixties cultural revolution, writes that the waning of the impulse to be a hippy was in part due to the fact that hippies' dominant values meshed with cultural changes that were affecting society at large, such that burned out hippies found it fairly easy to re-assimilate into "the straight world." He observes that "hippies were only a spectacular exaggeration of tendencies transforming the larger society" (Matusow, 1984, p. 306).
2. In comparison to earlier generations, sixties youth had an extraordinary amount of schooling and were insulated from the necessity of early entry into the workforce. They also had an expanded array of career opportunities. Far from relieving them of pressures relating to the selection and cultivation of identity components, these circumstances complicated the process by drawing it out and exacerbating the opportunity losses associated with a firm career decision.
3. This narrative is not meant to insinuate that parents in the 1960s were particularly rigid. Many were very tolerant and the largest concentrations of rebellious youth came from urban centers, especially in the Northeast and West Coast, where parents tended to be more liberal. Youth in the more conservative Midwest overwhelmingly stuck with the establishment. In fact, many of the instigators of the counterculture were not youth at all, but part of the parents' generation (O'Neill, 1971, p. 233).

4. Sandel rightly notes that any objection to the original position based on a psychosocial argument about the natural limitations of existing humans (e.g., to reason effectively under certain conditions, and so forth), is insufficient to defeat Rawls's project, which makes an epistemological rather than a psychological claim (Sandel, 1998, p. 12). The critical point is that, without importing some particular conception of the good that goes beyond what every individual is compelled by rationality alone to accept as good, Rawls simply has insufficient resources, epistemologically speaking, to justify his conception of justice as fairness.

5. As stated in the quoted passage, Daniels's reasoning is illogical, as his conclusion does not follow from, or for that matter even seem to be suggested by, his premise. But in view of the earlier observations about Rawls's metaphysics of the person, which presumably Daniels accepts, the conclusion seems less far flung (though still not required as a logical inference).

6. Recall that difficulty deciding between potential identities is a source of painful identity-deficit crises. Presumably the liability to this difficulty will increase as the inventory of potential identities is expanded. In other words, the pursuit of health, on Daniels's conception, may be injurious to health.

7. Even if it were possible to establish an ontology of persons apart from an account of natural psychology, Rawls's particular ontology would be susceptible to the usual objections about his neo-Kantian distinction between the right and the good, and his contention that the right is to some extent a priori (in the sense that a conception of justice can be articulated by ideal contractors who are disengaged from their empirical circumstances) and begets an a priori framework for identity formation, while the good arises a posteriori, under limitations imposed by the a priori right.

8. This may sound odd coming from a libertarian. But, though it is true that some versions of libertarianism do indeed enshrine individual choice, these are not the versions that I support. Furthermore, libertarians are less prone than liberal egalitarians to regard individual choice as the ontological basis for personhood, or even as the basis for social philosophy. Rather, they tend to view it as a political side constraint – recognizing a strong distinction between political philosophy (which respects the limits of moral knowledge) and social philosophy (which seeks as much as possible to overcome these limitations). Rawls, on the other hand, presumes and imports an austere, individualistic social philosophy, based on an ontology of choice, into his political philosophy. I have written on this topic extensively elsewhere (Trotter, 2001, 2002, 2003a), but will not treat it in detail here. The current essay does not contain an argument for libertarianism.

9. Brown's initial reconstruction of Freudian psychoanalytic theory was undertaken in his seminal book, *Life Against Death* (1959). Rejecting Freud's version of Eros-Thanatos as a biological duality afflicting the human subconscious, Brown regarded the human being as a essentially a creature of Eros, with Thanatos evoking a misguided flight from death. His prescription: accept death so that one can be resurrected, through polymorphic sexual experience, into a spiritual union of many bodies – thus fulfilling the Dionysian life impulse and achieving the authentic natural unity of the self.

10. There were, of course, many others. The story of Ken Kesey (Tom Wolfe, 1968) is exemplary. Subscribing to the credo of mystical unity through sex, drugs and rock "n" roll, Kesey regarded himself as a religious prophet destined to convert American into an "Electric Tibet." Matusow observes that Kesey's Trip's Festival in San Francisco provided a sense of social continuity to Haight-Ashbury's "acid freaks in search of identity" (Matusow, 1984, p. 292).

11. Kimball, in his section on "The triumph of babydom," quotes Yippie leader Jerry Rubin (yippies were a late-sixties concoction of Rubin and Abbie Hoffman – conceived essentially as hippies with a political agenda): "We're permanent adolescents" and "Satisfy our demands, and we've got twelve more. The more demands you satisfy, the more we got" (Kimball, 2000, pp. 9–10).

12. In the fifth grade, my oldest son was given a cherry-flavored condom by his sex education teacher, who told him that it might be fun to have a girl put it on his penis with her mouth. I did not find out about this until he was in high school – when he related that it was the only time in his life that he felt truly inadequate because he was not sexually active.

3 Genesis of a Totalizing Ideology: Bioethics' Inner Hippie

13. Thomson actually addresses the argument that a woman's voluntary engagement in sexual activity diminishes her rights over and against the fetus, but her rebuttal is so ridiculous that I cannot accept that she seriously considered the argument in the first place. Thomson's feeble new analogy goes as follows. There is a stuffy room and Thomson opens a window to air it, at which time a burglar climbs in. It would be absurd, she observes, for proponents of a burglar's right to her house and its contents to say that "she's given him a right to the use of her house – for she is partially responsible for his presence here, having voluntarily done what enabled him to get in, in full knowledge that there are such things as burglars, and that burglars burgle." This analogy breaks down on so many fronts that they are difficult to enumerate. Unlike the fetus, the burglar is himself guilty of a moral offense, he voluntarily enters the house, is not brought into existence by the voluntary action of the property owner, and so forth. A second analogy is more plausible, and gets at Thomson's fundamental moral conviction about sexual activity. In this analogy, there are "people-seeds" floating around in the air, and the woman who opens her window ends up with one in her house even though she fixes up her window with fine screens designed to keep the seeds out. Thomson claims that holding a woman using contraceptives responsible for pregnancy is as wrong as holding the woman in the analogue responsible for the seed that got in despite the screens. It just won't do, she opines, to expect people to live in stuffy quarters. Thomson doesn't mention that if contraceptives work well (and left liberals ordinarily insist that they work better than conservatives think), then most pregnancies where abortion is contemplated probably occur after sex without contraception. More to the point, she seems to think that it is as unreasonable to expect women who would kill their conceptus to abstain from sex as it is to expect them to live in stuffy homes. Sexual intercourse, she apparently thinks, is an essential component of healthy living – analogous to having good air. Either that, or she thinks that penises are so ubiquitous that it is as hard to keep them out of one's vagina as it is to keep pollens out of one's furniture.

14. I do not believe that there is hard evidence that smoking tobacco, in any form, is relatively safe in comparison to marijuana use. Though it is popular to claim that smoking marijuana increases lung cancer risk to about the same degree as smoking, there is thus far little evidence for this view. There is evidence that smoking marijuana, like tobacco, causes serious bullous lung disease in almost all cases of heavy use – and at a much faster rate than smoking tobacco, but such data do not prove that marijuana is equally dangerous overall. The "relatively safe" form of tobacco use to which I refer is dipping Swedish snus. In converting much of its population from smoking to snus use (which, unlike dipping American smokeless tobacco or smoking, is not associated with significant increases in the rate of oral cancer), Sweden has attained the lowest rates of lung cancer and oral cancer in Europe (Iceland excepted). Presumably activists in the European Union and World Health Organization who are eager to offer needles to IV drug addicts would also jump at the opportunity to offer snus to tobacco users. But that does not fly – tobacco companies are evil oppressors (unlike most marijuana distributors, who belong to an oppressed group – criminals). Hence, despite the likely loss of thousands of lives, the EU and WHO both oppose the legalization of snus in European countries where it is currently banned.

15. The transition from the rhetoric of love to the politics of rage parallels a similar transition in professional identity ideals I once described in the lives of some physicians – from the self-denying self to the rebel (Trotter, 1997, pp. 65–75). Royce considered this to be a fairly predictable intermediate phase in the evolution of personal or professional identities (Royce, 1967, 283–284). Nevertheless, what I have in mind in describing the New Left as a rebel movement is less concerned with eradicating inner demons, and (ironically given its roots) less concerned with detached individualism than the rebels Royce finds in Nietzsche and the stoics.

16. It would be difficult to articulate a more obviously false hypothesis. It is nonsense to think, for instance, that my health is directly imperiled if my neighbor gets a generous raise. My health might suffer if my neighbor advertises her good fortune, or spends her money ostentatiously

rather than investing it, saving it, or donating it to her church. In such a scenario the added risk to my health would be more proximally a result of factors that are only contingently related to income inequality – such as her spending habits and my tendency to react with jealousy, or to feel that I have less than I want or deserve. One possible contributor to the latter is an exposure to liberal egalitarian rhetoric that plays down the virtues of self-reliance, hard work and desert in favor of the rhetoric of individual entitlement. But of course, the bioethicists who champion the link between income inequality and ill health are themselves liberal egalitarians who hesitate to entertain, much less study such a hypothesis. They have commented, at times, about possible mediators of the linkage between inequality and ill health (Subramanian and Kawachi, 2004, p. 87), but income inequality remains an independent risk factor (for instance, in the way that smoking tobacco is an independent risk factor for lung cancer) only if these mediators are relatively fixed in all actual and all realistically possible societies, or if they vary proportionately and automatically with income inequality. Of course they do not. More sophisticated proponents of the thesis that economic inequality causes ill health have avoided claiming directly that the former is an independent risk factor for the latter, though this does not prevent them from assuming that it is when it comes to a discussion of "policy implications." They mention that correlation does not imply causality, throw out a few unsubstantiated hypotheses about causal linkages (always avoiding those that are incongruent with liberal egalitarian ideology), and then refer to economic inequality as a "social determinant" of poor health (Daniels et al., 2000). For most people, the terms "implication" and "determinant" reflect a very tight logical or causal linkage. Yet, interestingly, the European country with the least economic inequality (in terms of the heavily used GINI coefficient) is Denmark, which exhibits somewhat higher morbidity and lower life expectancy than other European countries with greater economic inequality (World Health Organization, 2008) – and rates a little lower on the good-health scale than the U.S.

17. The General Accounting Office in 1996 noted the existence of "higher claims costs in states with the most mandated benefits and more costly benefits, such as treatment for mental health and substance abuse," and observed that 39 mandates in Maryland (at that time the most in any state) accounted for about 22% of it claims costs, whereas Iowa's more modest mandates accounted for only 5% of its claims costs (General Accounting Office, 1996, pp. 8–9, 11). This problem has been exacerbated in many states in recent years. Primarily as a result of diverging government regulations, the average person in New Jersey pays $4,044 for one year of health insurance, while the average Iowan pays $1,188 (Matthews, 2005).

18. Also, 89% of U.S. households report that all household members "had enough food to eat" every day during the past year, and, of the remaining 11%, less than half were rated by the U.S.D.A. as experiencing "very low food security" – a category that involves some indication of reduced food intake, and that replaced "hunger" in 2006. Even among the small percentage of households that experienced very low food security, most did not have any individual who went a whole day without food, and the majority did not have an individual who lost weight (U.S. Department of Agriculture, 2007).

19. One cause of an inflated poverty rate is the reliance in computing it on the Census Bureau's Current Population Survey (CPS) for income data, and on the consumer price index (CPI) for adjusting the threshold for poverty over time. However, the CPS consistently under reports household economic resources (Rector et al., 1999), and the CPI overstated inflation by more than 1% for decades (Boskin et al., 1996) and by about 0.6% annually in the last few years, after improvements (Lebow and Rudd, 2003) – causing increasingly more misleading, and over-inflated, estimates of poverty. Also, welfare benefits have shifted in the directions of non-monetary awards and of monetary benefits (such as the Earned Income Tax Credit) that are not reflected in the calculation of the poverty rate. Finally, the government's determination of poverty does not take account of assets accumulated in prior years, such that some very well off households are included. The net result is that overall consumption (of food, space, energy, durable goods, etc.) in families below the poverty level, in comparison to reported incomes, has rapidly increased to levels above 200% (Eberstadt, 2006) and includes many

3 Genesis of a Totalizing Ideology: Bioethics' Inner Hippie 67

expensive items that can hardly be described as necessities. For instance, according to 2003–2005 data from the Census Bureau, 97% of poor households (as determined by the poverty level) own a color television, 55% have two or more color televisions, 63% have cable or satellite TV, and 43% own their own homes (Rector, 2007) – typically with about two rooms per person and about the same overall living space as middle class persons in Western Europe (Rector et al., 1999).

20. Card and Krueger (2005) argue that the link in classical economics between an increased minimum wage and increased prices does not pertain in the case of the fast-food industry because this market is not competitive, but rather exhibits elements of monopsony. I have seen numerous allusions to this study in the news media – usually inferring (wrongly) that Card and Krueger's findings would extend to other industries. They also generally fail to note that Card and Krueger's findings were overturned, and classical economic predictions vindicated, in a study by Neumark and Wascher (2005) that used payroll records, instead of the surveys employed by Card and Krueger, to estimate hours of employment. The debate rages on, but most economists persist in believing that increases in the minimum wage have deleterious effects on employment. Left liberal bioethicists would of course suspect that this convergence of opinion represents conservative bias. However, one of the surveys of economists (cited in the text of this essay in support of the aforementioned opinion) actually asked economists about their political party affiliations, and found that there were precisely four times as many Democrats as Republicans among respondents to the survey (Fuchs et al., 1998).

References

Bartlett, B. 2000. The verdict on the minimum wage: Guilty on all counts. *Economic Affairs* 20(3): 45–48.

Baumeister, R.F., J.P. Shapiro, and D.M. Tice. 1985. Two kinds of identity crisis. *Journal of Personality* 53(3): 407–424.

Beauchamp, T.L., and J.F. Childress. 2001. *Principles of biomedical ethics*, 5th ed. New York: Oxford University Press.

Boskin, M.J., E.R. Dulberger, R.J. Gordon, Z. Griliches, and D.W. Jorgenson. 1996. Final report to the U.S. Senate Finance Committee from the Advisory Commission to Study the Consumer Price Index.

Brooks, A.C. 2008. Liberal hatemongers. *Wall Street Journal*, January 17, Available: http://online.wsj.com.

Brown, N.O. 1959. *Life against death: The psychoanalytic meaning of history*. Middletown, CT: Wesleyan University Press.

Brown, N.O. 1966. *Love's body*. New York: Random House.

Burkhauser, R.V., and T.A. Finegan. 1989. The minimum wage and the poor: The end of a relationship. *Journal of Policy Analysis and Management* 8(1): 53–71.

Caplan, A. 2005. Abstinence-only sex ed defies common sense [On-line]. Available: http://www.msnbc.msn.com/id/9504871/print/1/displaymode/1098/.

Card, D., and A.B. Krueger. 1995. *Myth and measurement: The new economics of the minimum wage*. Princeton, NJ: Princeton University Press.

Carrington, W.J., and B.C. Fallick. 2001. Do some workers have minimum wage careers? *Monthly Labor Review* 124: 17–27.

Daniels, N. 1985. *Just health care*. New York: Cambridge University Press.

Daniels, N., B. Kennedy, and I. Kawachi. 2000. Justice is good for our health. In *Is inequality bad for our health?* eds. N. Daniels, B. Kennedy, and I. Kawachi, 3–33. Boston, MA: Beacon Press.

Eberstadt, N. 2006. *The mismeasure of poverty*. Washington, DC: American Enterprise Institute for Public Policy Research.

Friedman, T.L. 2003. Grapes of wrath. *New York Times*, March 12, Available: http://nytimes.com.

68 G. Trotter

Fuchs, V.R., A.B. Krueger, and J.M. Poterba. 1998. Economists views about parameters, values, and policies: Survey results in labor and public economics. *Journal of Economic Literature* 36(3): 1387–1425.

Gitlin, T. 1993. *The sixties: Years of hope, days of rage*. New York: Bantam.

Kelly, P.J., and L.R. Schwartz 2007. Abstinence-only programs as a violation of adolescents' reproductive rights. *International Journal of Health Services* 37(2): 321–331.

Kimball, R. 2000. *The long March: How the cultural revolution of the 1960s changed America*. San Francisco, CA: Enounter.

Lebow, D., and J. Rudd. 2003. Measurement error in the consumer price index: Where do we stand? *Journal of Economic Literature* 41(1): 159–201.

Levi, D.L., H. Stierlin, and R.J. Savard. 1972. Fathers and sons: The interlocking crises of integrity and identity. *Psychiatry* 35(1): 48–56.

Loewy, E.H. 2001. The social nexus of healthcare. *American Journal of Bioethics* 1(2): 37.

Matthews, M. 2005. Analysis: State mandates drive up insurance costs. *Health Care News*. Available: http://www.heartland.org/Article.cfn?artId=16867.

Matusow, A.J. 1984. *The unraveling of America: A history of liberalism in the 1960s*. New York: Harper and Row.

Nagel, T. 1975. Rawls on justice. In *Reading Rawls: Critical studies on Rawls' "A theory of justice,"* ed. N. Daniels, 1–15. New York: Basic Books.

Neumark, D., and W. Wascher. 2005. Minimum wage effects on employment and school enrollment. *Journal of Business & Economic Statistics* 13(2): 199–206.

O'Neill, W.L. 1971. *Coming apart: An informal history of America in the 1960s*. Chicago: Quadrangle.

Population Reference Bureau. 2008. Unmarried females who have had sex [On-line]. Available: http://www.prb.org/Datafinder/Topic/Bar.aspx?sort=v&order=d&variable=62

Rawls, J. 1971. *A theory of justice*. Cambridge, MA: Belknap Press of Harvard University Press.

Rector, R.E. 2007. *How poor are America's poor? Examining the "Plague" of poverty in America*. Washington, DC: Heritage Foundation.

Rector, R.E., K.A. Johnson, and S.E. Youssef. 1999. The extent of material hardship and poverty in the United States. *Review of Social Economy* 57(3): 351–387.

Royce, J. 1967. Royce's Urbana lectures: Lecture II, P. Fuss, ed. *Journal of the History of Philosophy* 5(3): 269–286.

Royce, J. 1995 [1908]. *The philosophy of loyalty*. Nashville: Vanderbilt University Press.

Sandel, M.J. 1998. *Liberalism and the limits of justice*, 2nd ed. New York: Cambridge University Press.

Subramanian, S.V., and Kawachi, I. 2004. Income inequality and health: What have we learned so far? *Epidemiologic Reviews* 26(1): 78–91.

Thomson, J.J. 1971. A defense of abortion. *Philosophy and Public Affairs* 1(1): 47–66.

Trotter, G. 1997. *The loyal physician: Roycean ethics and the practice of medicine*. Nashville: Vanderbilt University Press.

Trotter, G. 2001. Pragmatism, bioethics and the grand American social experiment. *American Journal of Bioethics* 1(4): W11–W30. Available: http://bioethics.net.

Trotter, G. 2002. Bioethics and healthcare reform: A Whig response to weak consensus. *Cambridge Quarterly of Healthcare Ethics* 11(1): 37–51.

Trotter, G. 2003a. Pragmatic bioethics and the big fat moral community. *Journal of Medicine and Philosophy* 28(5–6): 655–671.

Trotter, G. 2003b. Buffalo eyes: A take on the global HIV epidemic. *Cambridge Quarterly of Healthcare Ethics* 12(4): 434–443.

United States Census Bureau. 2005. Supplemental measures of material well-being: Basic needs, consumer durables, energy, and poverty, 1981 to 2002 [On-line]. Available: http://www.census.gov/prod/2005pubs/p23-202.pdf.

3 Genesis of a Totalizing Ideology: Bioethics' Inner Hippie

United States Department of Agriculture, E. R. S. 2007. Food security in the United States: Hunger and food security [On-line]. Available: http://www.ers.usda.gov/Briefing/FoodSecurity/labels. htm.

United States General Accounting Office. 1996. Health insurance regulation: Varying state requirements affect cost of insurance.

Whaples, R. 2006. Economist's voice (online journal): Do economists agree on anything? Yes! [On-line]. Available: http://www.bepress.com/ev.

Wolfe, T. 1968. *The electric kool-aid acid test*. New York: Farrar, Straus & Giroux.

World Health Organization. 2008. Highlights on health, Denmark 2004 [On-line]. Available: http://www.euro.sho.int/eprise/main/WHO/Progs/CHHDEN/home.

Chapter 4
Bioethics and Professional Medical Ethics: Mapping and Managing an Uneasy Relationship

Laurence B. McCullough

4.1 Introduction

Looking back on the development of the field of bioethics in the 1970s, one can now discern what was not so clear at the time: an uneasy relationship between bioethics and professional medical ethics, especially what Albert Jonsen (2000, 2009) has called the "long tradition" of medical ethics that preceded bioethics. On the one hand, there is a strand in early bioethics that treats bioethics as the application of ethical theory to problems in biomedical practice, research, and policy, a self-understanding of bioethics that deprofessionalized medical ethics. On the other hand, there is a strand in early bioethics that takes professional medical ethics seriously and includes professional medical ethics in the self-understanding of bioethics. In this chapter, I map this tension. In my judgment, the first strand has dominated the literature and its cutting-edge, reforming discourse continues to generate an excitement that draws students to undergraduate and graduate courses in bioethics in our colleges and universities. The second strand is less exciting, because it is conservative in that identifies what we should value in the history of medical ethics and therefore preserve and strengthen. This second strand properly shapes ethics teaching in our medical schools. I will propose to manage the uneasy relationship between bioethics and professional medical ethics, a problematic that we inherit from the early days of the field, by defending the second strand, by offering a defense of conservative, professional medical ethics.

L.B. McCullough (✉)
Dalton Tomlin Chair in Medical Ethics and Health Policy, Baylor College of Medicine,
Center for Medical Ethics and Health Policy, Houston, TX 77030-2411, USA
e-mail: mccullou@bcm.edu

H.T. Engelhardt, Jr. (ed.), *Bioethics Critically Reconsidered*,
Philosophy and Medicine 100, DOI 10.1007/978-94-007-2244-6_4,
© Springer Science+Business Media B.V. 2012

4.2 Bioethics that Deprofessionalized Medical Ethics

The *Encyclopedia of Bioethics* in its first edition in 1978 stands as one of the "monuments" of the field (Reich, 1978a). This is a term of art once used by art historians to designate paintings, drawings, prints, sculpture, and buildings that define the history of art and therefore constitute the "must-know" canon of the history of art. The *Encyclopedia* justly is one of the monuments of bioethics because, remarkably for a field then still young, it defined and consolidated bioethics and became its point of departure for the next two decades. The *Encyclopedia* is far more than a standard reference work, although it is surely that and one of the most remarkable of the genre. The *Encyclopedia* was meant to be, and remains, a reflection of the new field of bioethics, a rich mine marked by veins of the ores of self-understanding. I want to dig out some of those ores as exhibits to corroborate my claim that bioethics deprofessionalized medical ethics.

In the landmark entry, "Bioethics," in the *Encyclopedia*, K. Danner Clouser (1978), one of the founders of the field, succinctly summarized bioethics as "not a new set of principles of maneuvers, but the same old ethics being applied to a particular realm of concerns" (Clouser, 1978, p. 116). Taking medical ethics as "a first approximation" of bioethics, Clouser explains:

> Medical ethics is a special kind of ethics only insofar as it relates to a particular realm of facts and concerns and not because it embodies or appeals to some special moral principles or methodology. It is applied ethics. It consists of the same moral principles and rules that we would appeal to, and argue for, in ordinary circumstances. It is just that in medical ethics these familiar moral rules are being applied to situations peculiar to the medical world. We have only to scratch the surface of medical ethics and we break through to the issues of "standard" ethics as we have always known them (Clouser, 1978, p. 116).

The burning "issues" with which bioethics concerned itself were special, but the method for examining them was not. On this understanding of bioethics, professional medical ethics was distinctive only for its subject matter – medical practice and research – but not for its methods and core concepts. Moreover, the history of professional medical ethics was not pertinent to, part of, the self-understanding of bioethics. The ahistorical character of this first strand in bioethics reflected the then-prevalent ahistorical character of most of philosophy in the United States at the time.

It is worth noting that the *Encyclopedia*'s editor-in-chief, Warren Reich (1978b), took what, in retrospect, is the contrarian view that bioethics did indeed have a history that preceded the 1970s. The range of subject matter of bioethics should be understood to include "*historical perspectives*, particularly in the traditional area of medical ethics which deals specifically with the physician-patient relationship." (Reich, 1978b, p. xvii, emphasis original) To provide this needed historical perspective, Warren commissioned a 97,000-word series of entries on the history of medical ethics (Reich, 1978a, pp. 876–1007). "Medical Ethics: History of," presented a self-understanding of bioethics as including the history of medical ethics before bioethics. On this self-understanding bioethics was a recrudescence of medical ethics, with an expanded scope (Reich, 1978b, p. xix), rather than an invention.

4 Bioethics and Professional Medical Ethics: Mapping and Managing ... 73

But this was not the self-understanding that dominated bioethics at its "founding" four decades ago. Clouser's account is the more accurate.

Recently, Albert Jonsen (2000, 2009) has offered an account of the emergence of bioethics in the United States as a response to the perceived inadequacy of the "long tradition" of medical ethics that had preceded bioethics. Bioethics marked, as it were, a discontinuity, a break with the history of professional medical ethics. "Bioethics" "and the activity it designated, marks a boundary between the long tradition of medical ethics and a quite distinct approach to moral questions in medicine and science" (Jonsen, 2009, p. 477). Jonsen elaborates:

> At mid-twentieth century, this traditional medical morality encountered unprecedented problems. The conditions in which medicine was practiced had changed dramatically. Science had brought much more effective treatments. Many more persons had access to trained physicians and physicians became not only more educated and competent, but earned more money and social prestige than the profession had previously enjoyed. Old ideas about medical morality, particularly about the relationship with patients, were challenged by these new conditions (Jonsen, 2009, p. 478).

One of the principal inadequacies of traditional medical morality, paternalism, is underscored again and again in the early literature of bioethics, so much so that it almost became a rite of passage, an initiation ritual, to write on the topic of medical paternalism. In the *Encyclopedia* entry on paternalism, Tom Beauchamp (1978) defined it as "practices that restrict the liberties of individuals, without their consent, where the justification for such actions is either the prevention of harm they will do to themselves or the production of some benefit for them that they would not otherwise secure" (Beauchamp, 1978, p. 1194). Beauchamp goes on to add that "[t]here are so many individual examples of controversial paternalistic justification in biomedical and behavioral contexts that only a few selected examples can be treated here," (Beauchamp, 1978, p. 1197) and these were patient's refusal of treatment, human subjects research, telling the truth to patients, especially the seriously ill, behavioral control, biological control, and decision making for, rather than by, patients.

In the entry, "Code of Medical Ethics: Ethical Analysis," Robert Veatch (1978) describes what he characterizes as the "central ethic" of codes of medical ethics, a major genre of the history of professional medical ethics. This central ethic is that "the physician's first obligation to the sick person is to do what he thinks will benefit the sick person" (Veatch, 1978, p. 173). This "Hippocratic ethic" turns out to be a controversial ethic, on Veatch's analysis, not least because "the Hippocratic ethic is paternalistic" (Veatch, 1978, p. 173). On such an analysis, professional medical ethics is a problem to be solved and bioethicists like Veatch set out to solve the problem (see Veatch, 1981).

The self-understanding that emerges from these entries in the *Encyclopedia of Bioethics* is that bioethics is distinctive only in its subject matter, not its method. It method is the application of ethical theory, already thought to be adequate, to the distinctive subject matter. Ethical theory is generated from the moral point of view, characterized by an egalitarianism in which no one person counts more in moral decision making and behavior than another person. Special relations, e.g., of a parent to a child or of a physician to a patient, are discounted and explained by appeal to the concepts and discourse of ethical theory. Indeed, the application of the already accepted and adequate "general rules" of morality, as it were, explains away professional medical ethics. A textbook fresh off the press illustrates this nicely:

> The professional-patient relationship can be viewed essentially as one involving promises or contracts or, to use a term with fewer legalistic implications, covenants. The relationship is founded on implied and sometimes explicit promises (Veatch et al., 2010, p. 13).

In this view of professional medical ethics, contractually enforced promises are crucial for the regulation of relationships of unequal power. Implicit in this view of professional medical ethics is the assumption that unequal power is always predatory power of the party with the greater power. The person or entity with greater power vis-à-vis another and therefore with power over another is assumed not to be able to make reliable judgments about what is in the other's interests and, even if the person or entity could, that person or entity is relentlessly self-interested and will exploit the other for that person or entity's gain. The antidote to such incompetent, self-interested, and therefore predatory unequal power is anti-paternalism as the core meaning of the ethical principle of respect for patient autonomy.

> That is where autonomy surfaces as a *moral* principle. The moral principle of respect for autonomy holds that an action or practice is morally wrong insofar as it attempts to control the actions of substantially autonomous persons on the basis of coercion for their own welfare (Veatch et al., 2010, p. 115).

This strand in the self-understanding of bioethics deprofessionalizes medical ethics. By this I mean the discounting, even elimination, of fiduciary responsibility for patients, especially the obligation of physicians to protect patients from themselves, including patients who are undoubtedly competent but are making clinically unwise decisions. This can be called a paternalistic regard toward patients and sometimes may justify persuasion and even manipulation of patients who are autonomous. In a deprofessionalized medical ethics, the relationship is contractual and based on mutual exchange of promises. Implicit in such an account is the view that physicians, as the party with the greater unequal power, should not be understood to be committed to the care and protection of patients but as potential predators on them.

4.3 Bioethics that Embraced Professional Medical Ethics

An important component of bioethics from its inception has been health law. Indeed, court rulings and, later, statutes have been, and remain, enormously influential in and on the literature of bioethics. It is not well appreciated that health law has consistently embraced the centrality of professional medical ethics.

Perhaps the most notable examples come from the common law regarding end of life decision making. *In re Quinlan*, decided by the Supreme Court of New Jersey in 1976, surely counts as one of the monuments of bioethics (*In re Quinlan* 1976). This case concerned whether to grant the petition of Mr. Joseph Quinlan to be appointed guardian or the person and property for his unmarried, childless adult daughter. In the nomenclature that emerged from this case, he sought court approval to serve as his daughter's surrogate decision maker, because she was not competent to make her own decisions. This was because Ms. Karen Quinlan was determined to be in a "chronic, persistent vegetative state" (*In re Quinlan* 1976), what would now be called a permanent vegetative state.

Among the considerations addressed by the New Jersey Supreme Court was what it called the "medical factor," i.e., whether the decision to discontinue life-sustaining treatment for Ms. Quinlan "unwarrantably offends prevailing medical standards." Implicit in this formulation is the view that medicine is indeed a profession, that it has the autonomy to formulate and enforce standards of patient care. This autonomy presupposes that medicine is competent to set those standards and can be trusted to adhere to them conscientiously, i.e., by keeping the interests of patients paramount in clinical judgment, decision making, and behavior.

The Court formulated the issue as follows:

> The question is whether there is such internal consistency and rationality in the application of such standards as should warrant their constituting an *ineluctable bar* to the effectuation of substantive relief for plaintiff at the hands of the court (*In re Quinlan* 1976, emphasis added).

The Court then relied on a review of the medical literature on the limits of the obligation to treat dying patients that had appeared in the two preceding decades. The Court summarized the results of its analysis:

> We glean from the record here that physicians distinguish between curing the ill and comforting and easing the dying; that they refuse to treat the curable as if they were dying or ought to die; and that they have sometimes refused to treat the hopeless and dying as if they were curable (*In re Quinlan* 1976).

In short, the accepted standards of medical care set limits on the obligation to prolong life when the patient has an incurable, ultimately life-taking condition. There is therefore no "ineluctable bar" to Ms. Quinlan's physicians to accept her father's request that she be disconnected from the respirator that, among other measures, was keeping her alive.

Little noticed in the bioethics literature is the implication of the Court's line of reasoning. Had the Court's examination of the medical ethics literature resulted in the conclusion that there was an "internal consistency and rationality in the application" of a standard that required continuation of life-sustaining treatment of patients with incurable life-taking diseases, injuries, or conditions (i.e., if the Court had discovered that, contrary to the history of medical ethics, medicine is a vitalist profession), then the Court would have ruled against Mr. Quinlan's request on the basis of protecting the integrity of medicine as a profession. The concept of professional integrity cannot be generated from a contractual account of "professional" medical ethics based on mutual promise-making and -keeping.

4.4 The Invention of Professional Medical Ethics

The *Quinlan* Court did not identify when in the long tradition of medical ethics that preceded the 1950s (the farthest back that the Court looked in the history of medical ethics before bioethics (*In re Quinlan* 1976)) the concept of medicine as a profession and professional medical ethics were invented. This occurred at the end of the eighteenth century in Great Britain, in the work of the two giants of the modern period in the history of English-language medical ethics, the Scottish physician-ethicist, John Gregory (1724–1773), and the English physician-ethicist, Thomas Percival (1740–1804).

Gregory (1772) and Percival (1803) made their history-altering contributions to medical ethics in the context of and in response to nature of medical practice and research at the time (McCullough, 1998, 2009; Baker, 2009). In eighteenth-century Britain, and its North American colonies, medicine was very entrepreneurial. There was a deep tension within the Hippocratic texts between a life of service to the sick, on the one hand, and entrepreneurial self-interest, on the other. In the many centuries between the Hippocratic era and the early Enlightenments, this tension had been resolved in actual clinical practice outside faith communities in favor of the entrepreneurial, self-interested practice of medicine.

Porter and Porter (1989) provide a masterful account of the medical world of practitioners and the sick that existed in eighteenth-century Britain. There was a genuine marketplace of medical practice. Anyone could enter it as a practitioner, because there was no stable medical curriculum, no effective licensure (the efforts of the royal colleges to use their power of licensure to create monopoly control failed), no third-party payers, and no regulation of drugs and devices. There was no economic security for the considerable variety of practitioners, other than inherited wealth (Percival) or a handsome dowry (Gregory) for the lucky few. The market was supplied by university-trained physicians, apprentice-trained surgeons and apothecaries, female midwives, man-midwives (physicians who practiced obstetrics, using the new technology of forceps which they kept out of the hands of the female midwives), and many so-called irregulars. There were almost as many concepts of health and disease, and remedies for the latter, as there were practitioners.

The sick also routinely self-diagnosed and self-treated, which was known then as self-physicking. Many of the sick did so from economic necessity, because only the well-to-do could afford physicians' and surgeons' fees. Many of the sick self-physicked because they did not trust practitioners intellectually – to know that they were saying and doing – or morally – to put the interests of the sick first and their own financial and other self-interests second. The mistrust of the sick in medical practitioners was well founded.

This medical marketplace was very crowded and the number who could afford practitioners' fees few; the resulting competition therefore became fierce. The outcome of failure to compete successfully was poverty, again, with the exception of the handful of physicians who did not have to depend for their economic security on their fees. To gain competitive advantage, physicians did what they could to stand out, including adopting what Gregory (1772) called "little peculiarities" of dress, speech, and manners. It was essential to master the accoutrements of being a gentleman, to become socially acceptable to those with the wealth and therefore higher social standing who retained one's services. These were not mere matters of etiquette, i.e., mere manners, but the self-interested strategies of practitioners facing harsh and unforgiving market realties in which the well-to-do sick held and wielded the power of the purse over physicians.

The relationship between the sick and practitioners was contractual, with the well-to-do sick paying the piper and calling the tune (which payers for healthcare still do). As their payers, the well-to-do sick had considerable power over physicians; relatively speaking, physicians had no or only little power over the sick who could afford physicians' fees. The great shibboleth of deprofessionalizing bioethics, the systematically paternalistic physician, simply did not exist in the private practice of medicine.

The situation for the sick poor was different. Mainly, they self-physicked. When that failed, those who were qualified (see below) could be who were admitted to the newly created infirmaries, hospitals for the working sick poor. Risse (1986) has provided an indispensable account of these new health care organizations. These hospitals were created by the employers of what would come to known as the workplaces of the industrial revolution, such as cotton mills, coal fields, and ports, to provide free medical care for the men, women, and children whom they employed. The "trustees," as they were known provided an annual "subscription." Their combined subscriptions constituted the operating budget for the hospital. The trustees deliberately underfunded the hospital, to put pressure on its lay managers – physicians were not in charge – to control costs.

The infirmaries served mainly the worthy sick poor. The trustees appointed physicians to serve as "faculty" of infirmary. These physicians were not paid but gladly accepted – indeed, sought out – such appointments, because they brought with them the approval of the leading lights of the newly wealthy, and increasingly powerful, class. This approval could be, and was, used to build one's private practice (just as a "clinical" appointment to a medical school faculty is used today by private-practice physicians to build their practices).

The unworthy poor, the lazy and shiftless who were considered to have brought poverty on themselves, were not admitted to the infirmaries, for their own good. While this may strike some readers as odious, readers need to be aware that this distinction between the worthy and unworthy sick poor came across the Atlantic Ocean to the British American colonies and it with us still. In the United States we have Medicare and the Veterans Health Affairs system for the worthy sick poor and Medicaid, and city and county hospitals, for the unworthy sick poor. The political support for the former is strong, while for the latter it is weak, reflecting with precision eighteenth-century moral, social, and political categories.

The trustees had complex motives in their creation of and ongoing support for infirmaries. Chief among these was their interest in advancing their social and political standing vis-à-vis the landed aristocracy. As a consequence, the trustees wanted their infirmaries to look good and looking good meant low mortality rates. Physicians since the time of Hippocrates understood how to achieve this goal:

become very good at the prognosis of incurable, life-taking conditions so that one could in a timely, self-interested fashion declare a case incurable and withdraw. By abandoning the dying in such a timely fashion, incurable disease and not the physician would be likely blamed for the ensuing death. (It is still the case that patients in academic medical centers die only from incurable diseases, injuries, and conditions and not from anything that those caring for them do or fail to do.) Abandoning the incurable to their fates became the standard of practice and was codified into medical ethics by the early eighteenth-century German physician-ethicist Friedrich Hoffmann (1660–1742), in his *Medicus Politicus* (1749), the politic doctor (See Jonsen, 2000). The trustees accepted the standard of abandoning the incurable and instructed lay managers accordingly: deny admission to the sick with "fever, " which was a catchall diagnostic category of conditions with a high likelihood of morality even with treatment. Physicians were not to be entrusted with this important rationing task. The tactic worked. In a hospital with no concept much less science, and therefore no effective practice, of infection control, mortality was kept very low (Risse, 1986).

The trustees were very concerned about cost control, anticipating by two centuries the current context in which both private and public payers are vitally concerned about the cost of hospital care. The trustees did not trust physicians and surgeons with control of the most expensive resources in the hospital: the drugs, fortified wines, and beers in the formulary. These they put under the control of apothecaries, thus making physicians accountable to non-physicians, again anticipating one of the principal management tools of what came to known as managed care in the latter decades of the twentieth century.

In the infirmaries, physicians for the first time gained power over the sick, who unlike the paying sick, came from social classes beneath the modest social class of physicians and surgeons. The sick poor, who referred to themselves as "inmates," worried about the abuses of this power, e.g., for purposes of conducting experiments of their newly invented secret remedies or nostrums, an abuse that Gregory attacked (Gregory, 1772).

Gregory based his medical ethics on David Hume 's (2000) moral science and its major discovery, the principle of sympathy (McCullough, 1998). Sympathy constitutes our moral physiology: the natural capacity of each human being to enter into and experience the sufferings of others and to be motivated routinely to relieve and prevent such suffering. To be well functioning, sympathy needed to be regulated by virtues such as tenderness and steadiness, the exemplars of which for Gregory were women of "learning and virtue." Based on Humean sympathy, Gregory thought that the plight of the sick, both well to do and worthy poor, was not acceptable. Moreover, Gregory was very concerned that the entrepreneurial, self-interested practice of medicine introduced biases that disabled clinical judgment and decision making, calling into question the very competence of physicians. Gregory addressed the problem of competence by appealing to the philosophy of science and medicine of Francis Bacon (1561–1626). Bacon called for medicine to be based on "experience," i.e., the rigorously collected and described results of natural and designed experiments. In effect, Bacon called for medicine to become evidence-based, a call that medicine is finally answering, four centuries after it was issued.

Before Gregory, physicians routinely used the word "profession" to describe themselves. However, they used this word in a self-interested way, to distinguish themselves from practitioners who had not attended a university and received a "regular" education (a dubious claim on its face, given the absence of a stable medical curriculum) and that their competitors – surgeons, apothecaries, female midwives, and "irregular" practitioners – had not done so and were therefore inferior practitioners.

Gregory set out to give "profession" genuine intellectual and moral content, as an antidote to the self-interested use of the word. In doing so, he put in place the first two of the three components that constitute the ethical concept of medicine as a profession (McCullough, 2006).

The first component of the ethical concept of medicine as a profession and therefore of professional medical ethics is that physicians should become and remain scientifically and clinically competent. For Gregory, this meant that physicians should practice Baconian, experience-based medicine. In contemporary terms, this means that physicians should practice according the intellectual and clinical discipline of evidence-based medicine. When physicians routinely do so, they justifiably invite the sick to trust them intellectually, to know what they are saying and doing. Gregory was well aware that the sick did not trust their physicians intellectually and were usually justified in withholding their intellectual trust.

By becoming competent physicians gain intellectual authority for their clinical judgments about patients' health – its preservation through primary prevention and its restoration through secondary and tertiary prevention in response to disease and injury. In professional medical ethics the scientific and clinical competence of physicians is confined to health, which has come to be understood in biopsychosocial terms (Engel, 1977), an especially pertinent model to the emerging era of genomic medicine. Physicians set themselves up for entirely preventable ethical conflict when their fail to stay within the bounds of their professional expertise and authority. A project worth undertaking, but beyond the scope of this chapter, would be a study of the literature on medical paternalism, to distinguish alleged instances of paternalism about matters in which physicians have professional authority, those related to health, and matters in which physicians exceed their professional authority. The latter may be instances of paternalism, but not medical paternalism.

The second component of the ethical concept of medicine as a profession and therefore of professional medical ethics is that physicians should commit themselves to the protection and promotion of the health-related interests of the sick as their primary concern and motivation and keep self-interest systematically secondary. The second component of ethical concept of medicine as a profession requires considerable self-sacrifice. Percival was keenly aware of this, e.g., in his call to avoid consultations (except for emergency cases) on the Sabbath (Percival, 1803). The ethically justified limits of the professional virtue of self-sacrifice, especially sacrifice of income, convenience, and time of physicians and trainees, have become a central topic for professional medical ethics, but barely register in deprofessionalized bioethics.

The first two components of the ethical concept of medicine as profession are expressed in the professional virtue of integrity, a bedrock virtue of professional medical ethics. Without being aware of its historical origins in Gregory's medical ethics, the *Quinlan* Court correctly understood the centrality of the professional virtue of integrity. It obligates physicians to practice medicine, conduct research, and teach to standards of intellectual excellence (the first component of the concept of medicine as a profession) and to standards of moral excellence (the second component of the concept of medicine as a profession).

Gregory gestured in the direction of the third component of the ethical concept of medicine as a profession and therefore professional medical ethics when he excoriated the "corporation spirit" of the organized medicine of his own day, i.e., the royal colleges operating under the auspices of a royal charter (Gregory, 1772). The royal colleges were, essentially, merchant guilds that existed and conducted themselves primarily for the sake of the interests of their members in gaining and holding market share. The self-interested nature of the royal colleges was reflected in their *statuta moralia* (moral statutes) or rules of conduct. These included prohibitions against attacking brother physicians in public, so as not to injure the profession, i.e., injure the interests of guild members in presenting an attractive public face to their potential customers among the well to do (McCullough, 1998). The *statuta moralia* were self-interested and thus do indeed qualify as rules of mere etiquette among physicians.

Percival should be given the credit for picking up on Gregory's gesture and explicitly developing it conceptually, in his discussion of the ethics of when a physician or surgeon should retire from practice. He provides insightful analysis of the intellectual skills such as an acute memory and the ability to reason by analogy from present to past cases and, for surgeons, skills such as "quickness of eye-sight, delicacy of touch, and steadiness of hand, which are essential to the skilful performance of operations" (Percival, 1803, p. 52). This passage is especially moving in light of the fact that Percival by then had become blind and himself and retired from medical practice, thus following and exemplifying his own ethical analysis and the conclusions that it required. He concludes:

> Let both the physician and surgeon never forget, that their professions are public trusts, properly rendered lucrative whilst they fulfil them; but which they are bound, by honour and probity, to relinquish, as soon as they find themselves unequal to their adequate and faithful execution (Percival, 1803, p. 52).

Gregory and Percival wrote the first professional medical ethics in the history of Western medical ethics, perhaps in the global history of medical ethics. As a consequence, they became the first to use the word "patient" rather than the phrase "the sick" – for the Latin, *aegrotus*, which is the word used in previous texts and often mistranslated as "patient". In using "patient" as a moral category, Gregory and Percival drew an important implication from the ethical concept of medicine as a profession. Until medicine became a profession, a process that started with Gregory and Percival and still has a long way to go if critics such as Rothman (2000) are correct, there are no physicians, only practitioners of various kinds. Routinely fulfilling the three commitments required by the concept of medicine as a profession

4 Bioethics and Professional Medical Ethics: Mapping and Managing ...

turns medical practitioners into professional physicians. In other words, the profession of medicine is not a given that comes down to us robustly intact from the pen of Hippocrates. Rather, the profession of medicine exists as a function of the collective clinical judgments, decisions, and behaviors of physicians. External entities such as payers do not create the profession of medicine and they cannot destroy or injure it. Physicians are fully in charge of both and should hold themselves accountable for both.

When medical practitioners become professional physicians by committing to the three components of the ethical concept of medicine as a profession and therefore to professional medical ethics, the sick become patients. Patients can and should trust professional physicians both intellectually and morally. The older, contractual relationship of the sick and hired practitioners becomes replaced with a physician-patient relationship as a fiduciary relationship of protection and promotion of the patient's and research subject's health-related interests.

4.5 In Defense of a Conservative, Professional Medical Ethics

The strand in bioethics that deprofessionalizes medical ethics thinks of itself as progressive, reforming the rampant paternalism of physicians that this strand of bioethics, mistakenly, was widely practiced and normative acceptable (McCullough, 2011). Reformist bioethics focuses on the present and future, treating its agenda of controversial and engaging issues as new and unprecedented. By contrast the strand in bioethics that embraces professional medical ethics is conservative (the lower-case "c" should be emphasized; the comment is *not* political). The *Quinlan* Court, focusing on professional integrity as a moral and legal consideration that could be "ineluctable" focused on the past, seeking to conserve what is worth conserving. Professional medical ethics should be understood to be conservative: it seeks to conserve an important ethical concept from the history of medical ethics, medicine as a profession.

We should prefer conservative professional medical ethics over progressive, reforming bioethics, precisely because of the protections that a fiduciary relationship between physicians and patients affords to patients. This protection is not available in deprofessionalizing bioethics. The first strand of bioethics is ahistorical. When viewed critically from a stance in the professional medical ethics of Gregory and Percival, deprofessionalizing bioethics can be called to account on whether it affords to the sick the protection from predatory power that professional medical ethics afford to patients.

A crucial result of subjecting contemporary, deprofessionalizing bioethics to an historically informed critical assessment is that deprofessionalizing bioethics is not justified in using the words "physician" or "patient" or in using the phrases "physician-patient relationship" or "patient-physician relationship". These words and phrases take their meaning from professional medical ethics *and only from professional medical ethics*. When these words and phrases are disconnected from their historical source, they become a discourse that, in MacIntrye's (MacIntyre, 1984) terms, can be characterized as shards of the past. More to the point, this discourse becomes parasitic on the very professional medical ethics that deprofessionalizing bioethics long ago rejected as problematic in its paternalism and other affronts to the autonomy of persons.

Consider again the entry, "Codes of Medical Ethics: Ethical Analysis" from the first edition of the *Encyclopedia of Bioethics*. To repeat, in this entry Veatch characterizes what he calls the "central ethic" of Hippocratic, codified medical ethics: "the physician's first obligation to the sick person is to do what he thinks will benefit the sick person" (Veatch, 1978, p. 173). The Hippocratic Oath, however, was mainly a guild oath, designed to formalize the induction of men who were not sons of Asclepian physicians and who therefore could not be assumed to be loyal on the basis of blood kinship (Jouanna, 1999). Moreover, there was no continuous tradition of taking or swearing the Hippocratic Oath (Nutton, 2009), which calls into question the historical reliability that there was a Hippocratic, central ethic in the history of Western medical ethics. Hippocratic physicians were entrepreneurs and the relationship between medical practitioners and the sick was contractual. This became the ethical standard for their relationship until Gregory and Percival reformed it into a professional relationship based on professional medical ethics.

From an historically informed critical perspective the use by Veatch of "sick person" is correct, but "physician" is not. "Medical practitioner" should be used instead, because professional medical ethics had not yet been invented in ancient Greece. Moreover, so-called Hippocratic medical ethics was contractual. But this is precisely what is called for in deprofessionalizing medical ethics, which now must be seen as unwittingly recapitulating the history of Western medical ethics from the ancient world to the end of the eighteenth century. There is nothing new, unprecedented, or reformist at all about deprofessionalizing bioethics.

Once we make this seemingly small correction in language-use, we can appreciate that the anti-paternalism of deprofessionalizing medical ethics is misplaced. As clearly articulated by Beauchamp (1978), paternalism is the interference with a person's autonomy for the good of that person. Deprofessionalizing medical ethics has focused on the first half of the definition and therefore has neglected to understand fully the second half: one must be able to claim some competence and therefore authority for making the judgment for another person what is good for him or her as the justification for the interference with his or her liberty. In short, paternalism requires as a necessary condition the first component of the ethical concept of medicine as a profession. Without a commitment to scientific and clinical competence and the authority about health-related interests that such a commitment creates, no medical practitioner can plausibly claim to make judgments about what is in the health-related interests of a sick person. In a contractual, non-professional practice of medicine there cannot be any paternalism, although there can be the exercise of predatory power, i.e., power that interferes with liberty. (This brief reflection on medical paternalism helps us to understand one its key features in bioethics: the interference with liberty is the main, perhaps exclusive problem, not the claim to intellectual authority about health, disease, and injury.) (McCullough, 2011).

4 Bioethics and Professional Medical Ethics: Mapping and Managing ... 83

In the contractual, non-professional practice of medicine, there can occur the predatory behavior of just the sort that generated the crisis of moral trust that Gregory and Percival set out to correct. An historically informed critical assessment of contractual, deprofessionalizing bioethics obliges us to conclude that contractual bioethics should not be expected to protect the sick from the predatory power of medical practitioners, precisely because contractual medical ethics failed to do so in practice. In the name of protecting persons' autonomy from paternalistic physicians, deprofessionalizing bioethics puts the sick at risk of predatory power from which contracts will not protect them because contracts are merely contracts.

4.6 Conclusion

To gain protection from the predatory power of medical practitioners, the sick need to become patients. But the sick only become patients when medical practitioners become physicians, i.e., when medical practitioners commit themselves to the three components of the ethical concept of medicine as a profession, to conservative professional medical ethics that is designed precisely to eliminate predatory power. Gregory makes this abundantly clear when he writes: "Every man has a right to speak where his life or his health is concerned, and every man may suggest what he thinks may tend to save the life of his friend" (Gregory, 1772, p. 33). This is, perhaps the first occurrence of the discourse of patients' rights in the history of Western medical ethics and the rights of patients derive from the professional relationship of physicians to patients. Thus, about self-physicking, Gregory states: "If a patient is determined to try an improper or dangerous medicine, a physician should refuse his sanction, but he has no right to complain of his advice not being followed" (Gregory, 1772, pp. 33–34).

Deprofessonalizing bioethics has no effective response to the predatory power of medical practitioners. Deprofessionalizing bioethics fails to protect the sick against paternalism, because it undermines the necessary condition for medical paternalism to exist in the first place, and fails to protect the sick from the predatory power of medical practitioners, because it has an historically unfounded confidence in the effect of contracts to protect the rights of the sick.

Bioethics as professional medical ethics conserves what should be conserved from the history of medical ethics. Bioethics as deprofessionalizing medical ethics, despite its self-understanding as breaking with the past, retains from the past and valorizes what should now be rejected: pre-professional, contractual medical ethics. Bioethics that deprofessionalizes medical ethics thus unwittingly creates the conditions for reproducing the centuries-long crisis of intellectual and moral trust in the history of Western medicine, in response to which Gregory and Percival invented professional medical ethics. This is not progressive bioethics at all but a significant regression. The uneasy relationship between bioethics and professional medical ethics should be managed by abandoning deprofessionalizing bioethics in favor of conservative, professional medical ethics.

References

Baker, R.B. 2009. The discourses of practitioners in nineteenth- and twentieth-Century Britain and the United States. In *The Cambridge world history of medical ethics*, eds. R.B. Baker and L.B. McCullough, 446–464. New York: Cambridge University Press.

Beauchamp, T.L. 1978. Paternalism. In *Encyclopedia of bioethics*, ed. W.T. Reich, 1194–1201. New York: Macmillan.

Clouser, K.D. (1978). Bioethics. In *Encyclopedia of bioethics*, ed. W.T. Reich, 115–127. New York: Macmillan.

Engel G.L. 1977. The need for a new medical model: A challenge for biomedicine. *Science* 196: 129–136.

Gregory, G. 1772. *Lectures on the Duties and Qualifications of a Physician*. London: W. Strahan and T. Cadell. Reprinted in McCullough, L.B., ed. *John Gregory's writings on medical ethics and philosophy of medicine*, 161–245. Dordrecht: Kluwer.

Hoffmann, F. 1749. *Medicus Politicus, sive Regulae Prudentiae secundum quas Medicus Juvenis Studia sua et Vitae Rationem Dirigere Debet*. Geneva: Fratres de Tournes.

Hume, D. 2000. *A treatise of human nature*, eds. D.F. Norton and M.J. Norton. Oxford: Oxford University Press.

In re Quinlan. 1976. 70 N.J. 10; 355 A.2d 647; 1976 N.J. LEXIS 181; 79 A.L.R.3d 205.

Jonsen, A.R. 2000. *A short history of medical ethics*. New York: Oxford University Press.

Jonsen, A.R. 2009. The discourses of bioethics in the United States. In *The Cambridge world history of medical ethics*, eds. R.B. Baker and L.B. McCullough, 477–485. Cambridge and New York: Cambridge University Press.

Jouanna, J. 1999. *Hippocrates*. Baltimore, MD: Johns Hopkins University Press.

MacIntyre, A.C. 1984. *After virtue*, 2nd ed. South Bend, IN: University of Notre Dame Press.

McCullough, L.B. 1998. *John Gregory and the invention of professional medical ethics and the profession of medicine*. Dordrecht: Kluwer.

McCullough, L.B. 2006. The ethical concept of medicine as a profession: Its origins in modern medical ethics and implications for physicians. In *Lost virtue: Professional character development in medical education*, eds. N. Kenny and W. Shelton, 17–27. New York: Elsevier.

McCullough, L.B. 2009. The discourses of practitioners in eighteenth-century Britain. In *The Cambridge world history of medical ethics*, eds. R.B. Baker and L.B. McCullough, 403–413. New York: Cambridge University Press.

McCullough, L.B. 2011. Was bioethics founded on historical and conceptual mistakes about medical paternalism? *Bioethics* 25: 66–74.

Nutton, V. 2009. The discourses of europeans practitioners in the tradition of the hippocratic texts. In *The Cambridge world history of medical ethics*, eds. R.B. Baker and L.B. McCullough, 359–362. New York: Cambridge University Press.

Percival, T. 1803. *Medical ethics: Or a code of institutes and precepts, adapted to the professional conduct of physicians and surgeons*. London: J. Johnson & R. Bickerstaff.

Porter, D., and R. Porter. 1989. *Patient's progress: Doctors and doctoring in eighteenth-century England*. Stanford, CA: Stanford University Press.

Reich, W.T., ed. 1978a. *Encyclopedia of bioethics*. New York: Macmillan.

Reich, W.T. 1978b. Introduction. In *Encyclopedia of bioethics*, ed. W.T. Reich, xv–xxii. New York: Macmillan.

Risse, G.B. 1986. *Hospital life in enlightenment Scotland: Care and teaching at the royal infirmary of Edinburgh*. Cambridge: Cambridge University Press.

Rothman, D.J. 2000. Medical professionalism – Focusing on the real issues. *New England Journal of Medicine* 342: 1283–1286.

Veatch, R.M. 1978. Codes of medical ethics: Ethical analysis. In *Encyclopedia of bioethics,* ed. W.T. Reich, 172–180. New York: Macmillan.

Veatch, R.M. 1981. *A theory of medical ethics*. New York: Basic Books.

Veatch, R.M., A.M. Haddad, and D.C. English. 2010. *Case studies in biomedical ethics*. New York: Oxford University Press.

Chapter 5
Two Rival Understandings of Autonomy, Paternalism, and Bioethical Principlism

Aaron E. Hinkley

5.1 Introduction

Beauchamp and Childress' *Principles of Biomedical Ethics* is built around an opposition to medical paternalism, as well as around a crucial and fatal ambiguity regarding their primary principle of autonomy (Beauchamp and Childress, 1979). On the one hand, Beauchamp and Childress invoke Kant's views of autonomy to explain the force of their principle of autonomy. On the other hand, they regard their principle of autonomy to be directed to respecting what Kant would recognize as heteronomous choices, that is, immoral choices. Their principle of autonomy and the bioethics it endorses are framed in terms of the ethos criticized by Griffin Trotter (Chapter 3, this volume), the ethos of doing things "my very own way", which lies at the heart of the culture that produced bioethics (Trotter, 2011). Given their endorsement of heteronomous individualism, Beauchamp and Childress mean for physicians to respect patients' choices, even when decisions are made on the basis of inclinations, not rational decision-making. Their respect of heteronomous, autonomous choices sets their principle of autonomy at tension with the principles of beneficence, non-maleficence, and justice, which concern the non-liberty-directed best interests of the patient.

This paper explores the foundational tension between these two understandings of autonomy, as well as the collision between Beauchamp and Childress' commitment to autonomy and the best interests of patients. In doing so, this paper will pick up a theme, also, raised by Laurence McCullough (2011) in his essay – how the crusade against medical paternalism, which shaped and directed much of early bioethics, in particular the bioethics of Beauchamp and Childress, is at odds with Kant's commitment to respect autonomous moral agency. *Pace* Beauchamp and Childress, it would, then, appear that a Kantian regard for autonomous choice would lead to an endorsement of physicians' attempting to direct their patients'

A.E. Hinkley (✉)
Department of Philosophy, MS-14, Rice University, Houston, TX 77005, USA
e-mail: hinkley@rice.edu

H.T. Engelhardt, Jr. (ed.), *Bioethics Critically Reconsidered*,
Philosophy and Medicine 100, DOI 10.1007/978-94-007-2244-6_5,
© Springer Science+Business Media B.V. 2012

decisions away from the heteronomous autonomous choices to which Beauchamp and Childress' principle of autonomy gives primary consideration.

A Kantian account of autonomy supports a version of paternalism that, while not coextensive with medical paternalism, runs counter to a cardinal agenda of Beauchamp and Childress in their bioethical principlism, respecting the actual decisions of patients, whatever those decisions might happen to be. According to Kant, not all decisions are equal. A Kantian understanding of autonomy does not respect the decisions of individual persons, including all the medical decisions those individual persons might happen to make. The Kantian understanding of autonomy respects those decisions that conform to its standard of morality and rationality, the categorical imperative. Some decisions by individuals live up to the standard of morality; many other choices simply do not. In fact, when Beauchamp and Childress engage the term "autonomy", it lacks for them the crucial normative dimension of Kant's understanding, in falling far short of Kant's understanding of autonomy, which is integral not just to moral agency, but to the praiseworthy character of such agents. It amounts to Kantian autonomy's antipode, heteronomy. Beauchamp and Childress, in asking physicians to respect decisions of patients that involve being determined by their inclinations, are asking physicians to respect patient decisions that may violate duties of patients to themselves to preserve their life and health. The implications of this state of affairs for the founding and foundations of bioethics as endorsed by Beauchamp and Childress is that one is offered a bioethics framed by the heteronomously self-directed moral fashions of the 1960s and 1970s, which support a moral view radically at odds with acting with Kantian autonomy. The concept of autonomy in Beauchamp and Childress' bioethical principlism is not merely unclear; it contains within itself two distinct understandings of what actually constitutes autonomy that are fundamentally contradictory. A critical examination of this tension places much of what would be termed medical paternalism by Beauchamp and Childress in quite a different and, perhaps, positive light.

5.2 Medical Paternalism and Autonomy in Bioethics

Bioethics' beginnings have the stated goal of establishing a space for individual autonomy for patients in medical decision-making.[1] This emphasis on individual patient autonomy has as one of its historical grounds as a reaction against the paternalistic role that physicians and other health care professionals have traditionally played in medical decision-making. Given the forces of the times, one can appreciate why bioethics as a discipline (as distinct from medical ethics as codes of ethics to be adhered to by the community of physicians and other health care professionals) began as a reaction against the perceived paternalism of the medical profession as a whole (Childress, 1979; Gert and Culver, 1979; Buchanan, 1978).

Bioethics emerged as an academic and clinical discipline in the wake of a wide range of legal changes bearing on the practice of medicine, some of which undermined medicine's status as a quasi-guild and transformed into a trade (see,

5 Two Rival Understandings of Autonomy, Paternalism, and Bioethical Principlism

for example, *The United States of America, Appellants, v. The American Medical Association, A Corporation; The Medical Society of the District of Columbia, A Corporation; et al.*, 317 U.S. 519 (1943); and *American Medical Assoc. v. Federal Trade Commission*, 638 F.2d 443 (2d Cir. 1980), and others which in many jurisdictions erased the standing of the professional standard for disclosure in informed consent and replaced it with the reasonable-and-prudent-person standard, or objective standard (see, for example, *Backlund v. University of Washington* (975 P.2d 950) 1999). There was a wide-ranging reaction against the way in which physicians possessed the moral authority of an independent self-governed profession (Engelhardt, 1996).

As H. Tristram Engelhardt, Jr., observes in his essay "The Ordination of Bioethicists as Secular Moral Experts," "The mid-twentieth century in America saw this guild status of medicine set aside in favor of a view regarding medicine as a trade. Federal court decisions that applied antitrust laws to the medical profession stripped medicine of many of its guildlike self-regulatory privileges" (2002, p. 70). By removing the anti-trust exemption from the medical profession, the federal courts in the United States transformed medical patients into consumers of medical goods and services. As consumers, the relative status of the patients to their health care providers was elevated, in that patients became autonomous actors in a market who enter into an economic exchange, either directly or indirectly, with physicians. Physicians for their part were supposed to support patients in the decisions made about the goods and services they receive in their exchange. As medical consumers, they were by virtue of accepting treatment from physicians not submitting themselves to the authority of the physicians.

In the shadow of these changes, there existed a perception especially in the literature at the beginning stages of bioethics that medical paternalism constituted a significant problem to be overcome, and bioethics as a field was directed against paternalism as a *raison d'etre*.[2] Bioethicists took upon themselves the crusade of eliminating medical paternalism, which they found throughout the western medical tradition. Consider a representative view from bioethicists concerning medical paternalism in the classical bioethical literature. According to Allen Buchanan in his paper, "Medical Paternalism," "there is evidence to show that among physicians in this country the paternalist model is the dominant way of conceiving the physician-patient relationship" (1978, p. 370). Indeed, according to Albert R. Jonsen (2003), in his *The Birth of Bioethics,* "It became common to assert that physicians were paternalistic, or inclined to make decisions about the welfare of their patients that patients were entitled to make for themselves. A Hippocratic maxim advised that patients did best when 'under orders.' " (1998, p. 326.) In the entry on "Paternalism" in the *Encyclopedia of Bioethics,* Beauchamp writes:

> Paternalism is the intentional limitation of the autonomy of one person by another, where the person who limits autonomy justifies the action exclusively by the goal of helping the person whose autonomy is limited. . . . Following this definition, an act of paternalism overrides the value of respect for autonomy on some grounds of beneficence. Paternalism seizes decision-making authority by preventing persons from making or implementing their own decisions (1995, pp. 1914–1915).

Its concerns about medical paternalism focused bioethics on supporting the unhindered choice of patients.

As one can glean from Beauchamp's entry in *The Encyclopedia of Bioethics* on paternalism (1995), Beauchamp and James Childress hold that medical paternalism impedes the autonomous decision-making of individuals. In the sixth edition of *Principles of Biomedical Ethics*, they write, "we define 'paternalism' as *the intentional overriding of one person's preferences or actions by another person, where the person who overrides justifies this action by appeal to the goal of benefitting or of preventing or mitigating harm to the person whose preferences or actions are overridden*" (2009, p. 208). In order to make a more nuanced distinction between hard and soft forms of paternalism, they subsequently claim that this definition is normatively neutral. However, it remains clear that the cardinal concern for Beauchamp and Childress in opposing at least most forms of medical paternalism is that medical paternalism interferes with the unhindered choices of patients considered as individual persons.

In another piece from the early literature of bioethics, Bernard Gert and Charles Culver, in their essay, "The Justification of Paternalism" lay out the criteria for an agent acting paternalistically towards another person. They write:

> A is acting paternalistically toward S if and only if A's behavior (correctly) indicates that A *believes that*: (1) his action is for S's good; (2) he is qualified to act on S's behalf; (3) his action involves violating a moral rule (or will require him to do so) with regard to S; (4) S's good justifies him in acting on S's behalf independently of S's past, present, or immediately forthcoming (free, informed) consent; (5) S believes (perhaps falsely) that he (s) generally knows what is for his own good. From this account, it becomes clear that paternalistic behavior needs justification because it involves doing that which requires one to violate a moral rule; indeed in almost all cases, it involves an actual violation of a rule. The definition also indicates that one feature used in justifying a violation of a moral rule is not open to A when acting paternalistically – namely, the consent of S toward whom A is violating the rule. For if A has S's consent, A's behavior is no longer paternalistic (Gert and Culver, 1979, p. 2).

It is clear in Gert and Culver's definition of paternalism that patient autonomy is for them the fundamental issue. The achievement of autonomous consent overrides any possible violations of other moral rules by medical professions. For Gert and Culver, paternalism violates moral rules because it does not gain consent for actions undertaken, however beneficial they might otherwise be. The cardinal violation of moral rules envisioned by Gert and Culver (and other bioethicists) is when a medical professional acts on a patient, even for her benefit, without her autonomous informed consent.[3] There appears to be no standard of morality to which bioethics appeals that is higher than the autonomous informed consent of patients. Through the desire to overcome medical paternalism, patients' autonomy became central to bioethics, so much so that Gert and Culver's conception of moral rules in medicine cannot be made sense of apart from the centrality of informed, un-coerced permission by patients. The question remains, however, what exactly is meant by "autonomy". While Beauchamp and Childress invoke Kant and his arguments regarding the autonomy of moral agents, the next section shows what it is that Beauchamp and Childress mean when they invoke autonomy as the first principle of their biomedical ethics.

5.3 Autonomy in Bioethical Principlism

The first principle of biomedical ethics for Beauchamp and Childress is for medical professionals to respect the autonomy of patients. "[T]o respect autonomous agents" entails, according to Beauchamp and Childress, "acknowledg[ing] their right to hold views, to make choices, and to take actions based on their personal values and beliefs" (Beauchamp and Childress, 2009, p. 103). Physicians and other health care professionals ought to then allow patients the right to have their own views, however, idiosyncratic, and make determinations about the health care services they will receive on the basis of their views, values, and beliefs. Moreover, according to Beauchamp and Childress,

> Such respect involves respectful *action*, not merely a respectful *attitude*. It requires more than noninterference in others' personal affairs. It includes, in some contexts, building up or maintaining others' capacities for autonomous choice while helping to allay fears and other conditions that destroy or disrupt autonomous action (Beauchamp and Childress, 2009, p. 103).

This respect for the autonomy of patients, therefore, entails more than simply not interfering with the decision-making capacities, or abilities, of patients. Respecting the autonomy of others, for Beauchamp and Childress, requires positive action on the part of health care professionals in order to support and nurture the ability of patients to make their own decisions based upon their own views, values, and beliefs. According to Beauchamp and Childress,

> Respect, in this account, involves acknowledging the value and decision-making rights of person and enabling them to act autonomously, whereas disrespect for autonomy involves attitudes and actions that ignore, insult, demean, or are inattentive to others' rights of autonomous action (2009, p. 103).

For bioethics, respecting the autonomy of moral agents implies supporting patients in making their decisions on the basis of their own criteria, even when those criteria are wrong-headed. In particular, it requires that the medical professional not engage in attitudes and actions that impede the capacity of patients in making their own choices for the medical care they will receive, even if the physician is of the judgment that the patients will choose contrary to their own best interests.

Beauchamp and Childress, in exploring the principle of respect for autonomy, do not explore the range of legitimate manipulation. They maintain that health care professionals should allow individual patients on their own terms to make free and informed decisions about the care they will receive. Health care professionals are to override the individual's previous, current, or forthcoming decisions even in what physicians may rightly consider to be in that patient's best interest. The commitment to preserving the autonomous decision-making of patients sets side-constrains on the other three principles, beneficence, non-maleficence, and justice. The principle of autonomy might appear at first blush to have Kantian roots.

5.4 Kantian Autonomy: Why the "Free" Choices of Patients Can Be Heteronomous

In *Principles of Biomedical Ethics,* Beauchamp and Childress write,
Kant argued that respect for autonomy flows from the recognition that all persons have unconditional worth, each having the capacity to determine his or her own moral destiny. To violate a person's autonomy is to treat that person merely as a means; that is, in accordance with others' goals without regard to that person's own goals" (Beauchamp and Childress, 2009, p. 109).

All of this is roughly true for Kant's account of autonomy. What Beauchamp and Childress fail to note is that it is only the autonomous moral agent who will in fact respect the autonomy of other moral agents. Indeed, the most crucial failure on the part of Beauchamp and Childress in their discussion of Kant and autonomy is that they seemingly fail to recognize the fact that for Kant the notion of the autonomy of moral agents is as such normative. For Kant, there is a certain way that autonomous moral agents should choose to act; autonomous moral agents should not choose to act in any fashion they see fit but only in accordance with the dictates of moral rationality as such.

According to Kant, "all rational beings stand under the *law* that each of them is to treat himself and all others *never merely as means* but always *at the same time as ends in themselves*" (1996a, 82, 4, p. 43). Beauchamp and Childress draw upon this notion when they argue that moral agents must respect the autonomy of other moral agents. As previously noted, for Kant there is a normative dimension that must be fulfilled by moral agents in order for them to be considered autonomous. It is not a property they have purely in virtue of the fact that they are moral agents. Moral agents are, for Kant, capable of autonomy, but that does not imply that they are in fact autonomous.

For Kant, moral agents are autonomous only if they do not allow anything other than the moral law defined as pure practical reason to guide their decisions. It is in this sense that according to Kant in *The Groundwork of the Metaphysics of Morals,* the autonomous moral agent is "subject *only to laws given by himself but still universal* and that he is bound only to act in conformity with his own will, which, however, in accordance with nature's end is a will giving universal law" (1996a, 82, 4, pp. 432–33). This moral law applies to all moral agents though only autonomous moral agents will live in accordance with this law. Kant further argues:

Autonomy of the will is the property of the will by which it is a law to itself (independently of any property of the objects of volition). The principle of autonomy is, therefore: to choose only in such a way that the maxims of your choice are also include as universal law in the same volition (1996a, 89, 4, p. 440).

For Kant, autonomous moral agents must at the same time in the same respect will the universal law as they are willing the maxim that can be abstracted from the principle of their action. Autonomy means willing the universal law when one wills one's action. This principle of moral action is what Kant names the categorical imperative.

5 Two Rival Understandings of Autonomy, Paternalism, and Bioethical Principlism

Moral agents, who do not act as moral lawgivers to themselves are understood by Kant to be heteronomous. They do not stand as lawgivers unto themselves because they make their decisions on the basis of something other than pure practical reason. Kant writes:

> For, if one thought of him only as a subject to a law (whatever it may be), this law had to carry with it some interest by way of attraction or constraint, since it did not as a law arise from *his* will; in order to conform with the law, his will had instead to be constrained by *something else* to act a certain way. By this quite necessary consequence, however, all the labor to find a supreme ground of duty was irretrievably lost. For, one never arrived at a duty but instead of the necessity of an action from a certain interest. This might be one's own or another's interest. But then the imperative had to turn out always conditional and could not be fit for a moral command. I will call this basic principle, the principle of the autonomy of the will in contrast with every other, which I accordingly count as heteronomy (1996a, 82–3, 4, pp. 432–33).

As heteronomous persons, these failed moral agents allow their decisions to be determined not by rationality as such, but by their own inclinations, feelings of attraction or revulsion, and not their rational will. They do things their way, not *qua* rational agent, but *qua* particular persons with particular inclinations. Kant argues that on the basis of inclinations, whether they are base or noble, one could never arrive at the universal law because such inclinations or interests might be one's own or someone else's, but they are diverse and, also, subject to change due either to a change in the object itself or a change in one's inclinations. According to Kant, "If the will seeks the law that is to determine it *anywhere else* than in the fitness of its maxims for its own giving of universal law – consequently, if, in going beyond itself, it seeks this law in a property of any of its objects – *heteronomy* alway result" (1996a, 89, 4, p. 441). The universal law of morality is to be found in practical reason as such, not in the objects, or properties of those objects, about which it makes judgments. It is clear from Kant's account of autonomy that autonomy is an essentially normative concept. Being a moral agent as such does not insure that one is autonomous. Moral agents who are autonomous will make decisions in accordance with the categorical imperative.

5.5 Kantian Autonomy as a Basis for Medical Paternalism

Unlike Beauchamp and Childress, for Kant there is a correct way for a person to choose in order to be autonomous. The autonomous persons are not simply persons capable of choosing and executing their decisions, but persons who choose the correct way in accordance with the categorical imperative and within the structures of practical reason. It is possible then in Kant's account of autonomy to know in a very strong sense what is best for another moral agent (even though in Kant's account each moral agent acting autonomously is a lawgiver unto herself but that is only so on the basis of practical rationality which is universal with all moral agents at all times being capable in principle of accessing it). For Kant, it is possible that one moral agent, who is acting autonomously, can know what the moral obligations of another genuinely are (e.g., accepting medical care so as to discharge duties to oneself to preserve life), so that the one who is autonomous should direct and guide the one in the grip of inclinations. In the case of medicine, physicians may recognize that patients are distorted in their medical judgments and need a physician's guidance. Of course, Kantian physicians should not lie to patients, but it would follow that they should attempt to direct patients to truly autonomous choices. The heteronomous decisions of patients would not deserve respect. Nor would attempts to bring them to autonomous choices be using them as means merely, but the very opposite.

This Kantian understanding of autonomy is in sharp contrast to what bioethicists, such as Beauchamp and Childress hold. In a very real sense, Kant, the philosopher upon whom Beauchamp and Childress draw to establish the principled priority of respect for autonomy, can be reasonably interpreted as giving grounds for a version of medical paternalism. Kantian paternalism in some regards is distinct from the medical paternalism that is the *bete noir* of bioethics in excluding the telling of any lies to patients, but it is nonetheless a version that may be even more foundational and apodictic than other forms of medical paternalism. This is because Kant would argue from what he holds is a synthetic a priori truth about moral agency, whereas any claim within medicine is only true as an a posteriori empirical truth. That the very philosopher Beauchamp and Childress invoke for respect of patient autonomy turns out to warrant a paternalism that supports much of the medical paternalism that Beauchamp and Childress decry is a supremely ironic fact of the matter.[4]

This is the case because Kant respects moral agents *qua* agents, as moral agents able to make autonomous choices, he is far from committed to respecting the heteronomous choices of moral agents and instead regards them as vicious. Moreover, what is clear is that an autonomous moral agent would not encourage other moral agents to make heteronomous choices, but the very opposite. While it is true that the autonomous moral agent for Kant respects the ability of other moral agents to make choices, it does not follow that Kant asks us to respect heteronomous choices or to refrain from encouraging others not to make heteronomous choices. However, in Beauchamp and Childress' account of autonomy, there is no requirement of independence from desires and inclinations, as there is for Kant, but only that the decisions of patients be made without coercion from others.[5] In Beauchamp and Childress' bioethical principlism, a moral agent can be considered autonomous, while in Kantian terms the agent is acting heteronomously, thus warranting paternalistic guidance.

5.6 Conclusion

At the very foundations of the establishment of bioethics, in particular that bioethics rooted in the reflections of Beauchamp and Childress, there is a fundamental confusion regarding the nature and importance of autonomy, as well as regarding

5 Two Rival Understandings of Autonomy, Paternalism, and Bioethical Principlism

why medical paternalism should be criticized. On the one hand, Beauchamp and Childress invoke Kant as a warrant for their account of their principle of autonomy and therefore for why physicians should respect the decisions of patients. They do this without recognizing that the autonomy which Kant calls persons in general and physicians in particular to respect is the autonomy of giving the moral law to oneself. Kant does not call persons in general and physicians in particular to respect, but at best only to tolerate the heteronomous, inclination-driven choices of others. On the other hand, Beauchamp and Childress equate autonomy with "doing it my way" not as a moral agent would do so in acting with Kantian autonomy, but according to whatever inclinations the agent embraces and affirms. Thus, the autonomy endorsed by Beauchamp and Childress' principle of autonomy is at loggerheads with Kant's account of autonomy, although Beauchamp and Childress invoke Kant's account as a warrant for the plausibility of their principle. There are two rival accounts of autonomy at the roots of Beauchamp and Childress's principle of autonomy.

It seems as if Beauchamp and Childress failed to notice the foundational difference in the meaning between the autonomy that they embrace and the autonomy that Kant endorses. In part, they may not recognize this difference because a Kantian notion of autonomy would surely warrant physicians in reproving their patients for making decisions contrary to the patient's obligation to preserve the patient's life and health. Surely, physicians guided by a Kantian moral perspective would be forbidden to lie to their patients or to coerce them in the sense of threatening force against them. But physicians inspired in a Kantian fashion would surely be warranted in attempting to bring their patients to appreciate the wrongheadedness of decisions that put their life and health at stake, thus violating duties of patients to themselves. For Beauchamp and Childress, the physician ought to respect the freely made decisions of their patients regardless of the reasons, or lack thereof, that the patient has for making their choice. However, given the Kantian understanding of autonomy, all choices are not equal; not all choices qualify as autonomous. Respecting the autonomy of the patient, from a Kantian understanding, might well call for some form of medical paternalism where the physician encourages the patient to make the rational, autonomous choice. That is, Kantian physicians would be at liberty, if not obliged, to instruct their patients regarding the immorality of smoking, a diet high in saturated fats, or of high-risk sexual relations, as well as to inform patients that they should accept those forms of treatment that are most likely to support the patients' obligations to preserve life and health. A Kantian understanding of autonomy would support physicians acting in ways that many would hold to be paternalistic in not accepting inclination-driven choices of patients as properly autonomous. Given the animus against medical paternalism in the 1960s and 1970s, one can appreciate how the cultural framework of the time shaped the early character of bioethics in embracing a view of autonomy that was a form of Kantian heteronomy. Now in critical retrospect, one can appreciate the incoherence of these reflections regarding autonomy at the very roots of bioethics and the implausibility of many of its claims.

Notes

1. Indeed, Beauchamp and Childress devote much space in the first edition of *The Principles of Biomedical Ethics* to the issue of medical paternalism. In each of the five subsequent editions, the issue of paternalism is dealt with in depth. Other examples of the concern about medical paternalism in the early stages of bioethics include Childress' essay "Paternalism and Health Care" and Gert and Culver's "Justification of Paternalism," as well as Allen Buchanan's "Medical Paternalism" (Childress, 1979; Gert and Culver, 1979; Buchanan, 1978). In establishing itself, bioethics consciously saw its *raison d'être* as curbing the influence of the paternalism of health care professionals in medical decision-making.
2. There remains the perception by those in the field of bioethics that medical professionals act paternalistically towards their patients and that this is a problem that bioethicists as a profession need to solve. In the entry on "Paternalism" on the *Stanford Encyclopedia of Philosophy,* a web based philosophical encyclopedia, as recently as 2005, Gerald Dworkin writes, "Doctors do not tell their patients the truth about their medical condition." The perception remains especially among bioethicists that health care professionals act paternalistically towards their patients, and that this is one of the many problems that bioethicists are needed to solve. Given its foundational role in the very existence of bioethics, it makes perfect sense that bioethicists would still perceive medical paternalism as a problem in need of a solution. Without medical paternalism, bioethics would have one less reason to exist as a discipline, their livelihood depends upon the continued perception of the problem. This is not even to answer the more fundamental question raised by Laurence McCullough's paper in this volume, "Bioethics' Ideology of Anti-Paternalism" of whether the perception of paternalism on the part of the medical community by bioethicists reflected the reality at any point in the history of Western medicine. Indeed, it is quite reasonable to ask: was the medical profession ever as paternalistic as the field of bioethics seems to have believed?
3. In his essay, "Paternalism and Health Care" Jim Childress disagrees with some of the specifics of Gert and Culver's definition of medical paternalism. He writes, "[T]he physician's action is paternalistic because his benevolent motives and intentions involve a refusal to accept the individual's own choices and wishes, not because it would lead to a violation of [a] moral rule" (1979, p. 20). Ultimately, however, the fact remains that bioethicists, especially in the beginning stages of the field, understood their criticism of medical paternalism to be their reason for existing as a field.
4. In similar vein Engelhardt points out the unintended consequences of bioethics' concern about paternalism and respect for patient autonomy in his essay "The Ordination of Bioethics as Secular Moral Experts". He writes, "Bioethicists have successfully assumed the role of conceptive ideologists, the role of those who can make "the perfecting of the illusion of the [ruling] class about itself their chief source of livelihood." In this role, bioethicists function not simply to frame a vision of appropriate deportment, but are politically useful in gaining control over a powerful, socially established domain of human intervention. Bioethicists have been socially ordained as being in authority to give expert guidance regarding the appropriate character of the third spatialization of disease, the social experience of human illness, disability, and defect. Bioethicists have thus become the custodians of the sociopolitical context that frames the world of health-care delivery. It is this establishment of applied moral philosophy in health care that constituted the social authorization or baptism of bioethics" (2002, p. 82). The irony here is that bioethical concern for patient autonomy, which began in part as a reaction against the paternalism of the medical profession has in the field of bioethics established itself as a moral-medical paternalism. Bioethicists have set themselves up as experts in morality as it relates to the field of health care. This new paternalism has its foundations in the variety of individualism taken as an un-argued for presupposition by the beginning stages of bioethics. By adopting in some sense the language of Kantian morality and its respect for individual, autonomous, rational, moral agents, bioethics took into itself a view that does not hold that one ought to respect individuals and their actual moral choices, but instead individual moral agents when they are acting in accordance to the dictates of rationality as defined by the Kantian philosopher. Individuals

5 Two Rival Understandings of Autonomy, Paternalism, and Bioethical Principlism

are only successful moral agents when acting as one ought to act according to the categorical imperative. If the individual fails to act in such a manner, they are acting heteronomously, not autonomously, and it is only their autonomy which one as a rational agent is bound to respect. Whereas medical professionals behaved paternalistically on the basis on their practical, medical expertise, bioethicists have allowed themselves to be paternalists of a more insidious variety on the basis of philosophical arguments. However, the question of the paternalism of bioethics as such is outside the scope of this paper.

5. Kant writes in *The Critique of Practical Reason*, "That *independence,* however, is freedom in the *negative* sense, whereas this *lawgiving of its own* on the part of pure and, as such, practical reason in the *positive* sense. Thus the moral law expresses nothing other than the *autonomy* of pure practical reason, that is, freedom, and this is itself the formal condition of all maxims, under which alone they can accord with the supreme practical law" (1996b, 166, 5, p. 33). Therefore, Kant himself recognizes before the fact that differing accounts of moral autonomy turn on the nature of independence from external forces. Beauchamp and Childress have what according to Kant is a negative understanding of independence, unlike Kant's own, which he claims is positive. Differing conceptions of autonomy as we have previously suggested would not necessarily be a problem for bioethical principlism except that Beauchamp and Childress draw heavily upon Kant in their principles.

References

American Medical Assoc. v. Federal Trade Commission, 638 F.2d 443 (2d Cir. 1980).

Backlund v. University of Washington (975 P.2d950) 1999.

Beauchamp, T. 1995. Paternalism. In *Encyclopedia of bioethics,* 2nd ed., ed. W.T. Reich, 1914–1920. New York: Macmillan.

Beauchamp, T., and J. Childress. 1979. *Principles of biomedical ethics.* New York: Oxford University Press.

Beauchamp, T., and J. Childress. 2009. *Principles of biomedical ethics,* 6th ed. New York: Oxford University Press.

Buchanan, A. 1978. Medical paternalism. *Philosophy and Public Affairs* 7: 370–390.

Childress, J. 1979. Paternalism and health care. In *Medical responsibility: Paternalism, informed, and euthanasia,* eds. W. Robison and M. Prichard, 1–14. Clifton, NJ: Humana Press.

Dworkin, G. 2005. Paternalism. In *Stanford encyclopedia of philosophy.* http://plato.stanford.edu/entries/paternalism/. Accessed 26 Jan 2010.

Engelhardt, H.T., Jr. 1996. *The foundation of bioethics,* 2nd ed. New York: Oxford University Press.

Engelhardt, H.T., Jr. 2002. The ordination of bioethicists as secular moral experts. *Social Philosophy & Policy* 19(2): 59–82.

Gert, B., and C. Culver. 1979. The justification of paternalism. In *Medical responsibility: Paternalism, informed, and euthanasia,* eds. W. Robison and M. Prichard, 15–28. Clifton, NJ: Humana Press.

Jonsen, A. 2003. *The birth of bioethics.* New York: Oxford University Press.

Kant, I. 1996a. The groundwork of the metaphysics of morals. In *Practical philosophy,* trans. M. Gregor, and ed. A. Wood, 37–117. New York: Cambridge University Press.

Kant, I. 1996b. The critique of practical reason. In *Practical philosophy,* trans. M. Gregor, and ed. A. Wood, 137–272. New York: Cambridge University Press.

McCullough, L. 2011. 'Bioethics' ideology of anti-paternalism. In *Bioethics critically reconsidered: Having second thoughts,* ed. H.T. Engelhardt, Jr., 71–84. Dordrecht: Springer.

The United States of America, Appellants, v. The American Medical Association, A Corporation; The Medical Society of the District of Columbia, A Corporation; et al., 317 U.S. 519 (1943).

Trotter, G. 2011. Genesis of a totalizing ideology: Bioethics' inner hippie. In *Bioethics critically reconsidered: Having second thoughts,* ed. H.T. Engelhardt, Jr., 49–69. Dordrecht: Springer.

Part II
The Practice of Bioethics and Clinical Ethics Consultation: Three Views

Chapter 6
Bioethics as Political Ideology

Mark J. Cherry

> *Ideology: 4. A systematic scheme of ideas, usu. relating to politics or society, or to the conduct of a class or group, and regarded as justifying actions, esp. one that is held implicitly or adopted as a whole and maintained regardless of the course of events. ... 1970 D.D. Raphael Probl. Pol. Philos. i. 17. Ideology... is usually taken to mean, a prescriptive doctrine that is not supported by rational argument (Oxford English Dictionary, On-line edition, 2008).*

6.1 Introduction

In their critical retrospective on American bioethics, "Examining American Bioethics: Its Problems and Prospects", Renée Fox and Judith Swazey diagnose numerous challenges they believe face the future of bioethics. They raise broad criticisms from the "secondary status accorded to virtues and qualities of the heart, like compassion, sympathy, kindness, and caring...", to what they perceive as the "field's relative inattention to ethical questions of social justice in American society..." and the "maintenance of a distinction between bioethical and human rights issues and of a line of demarcation between bioethics and international human rights law..." (2005, p. 363). Their conclusions include what they term an "admonitory statement" to the field of American bioethics and its role in the culture wars:

> The rancorousness has been conspicuous in statements made by bioethicists on both sides of the divide in professional publications, blog sites, and the media about matters such as embryonic stem cell research, therapeutic and reproductive cloning, genetic enhancement, other uses of fetal tissues and embryos, end-of-life issues, the place of religious thought and belief in ethical analysis, and about the potential dangers and harms as well as benefits that advances in medical science and technology can bring in their wake (2005, p. 369).

M.J. Cherry (✉)
The Dr. Patricia A. Hayes Professor in Applied Ethics, Professor of Philosophy,
Department of Philosophy, St. Edward's University, Austin, TX 78704, USA
e-mail: markc@stedwards.edu

H.T. Engelhardt, Jr. (ed.), *Bioethics Critically Reconsidered,*
Philosophy and Medicine 100, DOI 10.1007/978-94-007-2244-6_6,
© Springer Science+Business Media B.V. 2012

Fox and Swazey argue that such animosity "runs counter to the composed, fair-minded, rationally thoughtful outlook ideally associated with a philosophically oriented field" and they consider such strife as "... amoral, destructive, and self-destructive" (p. 369).[1] Their hope is that the field of bioethics will find a way to extricate itself from the culture wars.[2] In this chapter, I argue in support of this conclusion that bioethics should extricate itself from the culture wars. I conclude, however, that Fox and Swazey, and most others who share this same goal, do not seem fully to appreciate either (1) the substantial nature of such a conclusion or (2) the significant social and political implications of carrying out such a challenge.

6.2 The Public Ideology of Bioethics

If bioethics is to extricate itself from the culture wars, the field must alter many of its most fundamental characteristics. Instead of creating social space for the robust expression of diverse moral, religious, and secular worldviews, for "composed" and "fair-minded" rational dialogue, bioethics routinely seeks top-down governmental regulation, ideologically driven towards a fully secular vision of ethics and social justice.[3] There is a deep consanguinity between the development of bioethics as a field of inquiry and the ideological political movement to establish at law patient rights, welfare entitlements, and a secular social-democratic political vision (Engelhardt, 2009, p. 296). Fox and Swazey illustrate just such a political vision when they lament that bioethics draws what they perceive as too strong "... of a distinction between bioethical and human rights issues and a line of demarcation between bioethics and international human rights law..." (p. 363). It is unfortunate, they argue, that bioethics has not more fully embraced the international human rights legal framework to enforce this particular vision of social justice. Core human rights, such as rights to self-determination, and central values, such as personal autonomy and redistributive social justice, are believed to transcend regional, cultural, and religious differences, and thereby to bind all persons into a common moral framework. Human rights, it is claimed, authorize governmental authority to constrain and direct the medical, social, and bioethical choices of individuals, families and religious communities.

6.2.1 Example I: Human Rights and the Deconstruction of the Family

Consider, for example, the efforts in contemporary Western bioethics to limit the authority of parents in pediatric decision-making. There exists a widespread underlying assumption of human rights advocates that even minor children ought to be treated as self-possessed moral agents, who are to undertake their own moral and life-style decision-making as soon as possible and as far as feasible. In pediatric

6 Bioethics as Political Ideology

medical decision-making, emphasis is placed on protecting the child's "best interests", which themselves are typically appreciated in terms of the child's equality and liberty interests. Cardinal moral value is assigned to individual liberty conceptualized as autonomous self-determination. Persons as rational beings are to choose to be autonomous self-determining individuals, who shape moral values and perceptions of the good life for themselves. Individual autonomy is thereby highlighted as integral to human good and human flourishing. Parental authority is re-conceptualized in pediatric decision-making to give priority to the child's own self-determination, self-realization, individual equality, and actual or potential autonomy.

In its Convention on the Rights of the Child (September 02, 1990), the United Nations inserted itself into this moral and political debate, articulating an ethical and political vision that seeks to drive a legal and moral wedge between parents and their children. Throughout the Convention, the individual is prioritized as the locus of moral authority and primary unit of social value, where the celebration of free individualistic assessment of one's own moral values and life-style choices is accented as integral to the good life for persons. The Convention announces, without argument, an array of human rights for children, each of which emphasizes individual liberty and personal autonomy. Of particular interest, the Convention advanced the claim that children possess basic human rights to freedoms of expression, to receive information of all kinds (Article 13), of thought, conscience, and religion (Article 14), to privacy (Article 16, para 1), to freedoms of association and assembly (Article 15, para 1), as well as to free and compulsory human rights education (Article 29).

These basic human rights, the authors of the Convention judged, children hold independently of their parents and are enforceable by children over against their parents through direct appeal by the child to the state. As Richard Reading et al. endorse the implications:

> One of the far-reaching consequences of the UNCRC [United Nations Convention of the Rights of the Child] is that it makes the child an individual with rights and not just a passive recipient, and hence the child has the right to actively participate at all levels of decision making. The traditional association between the state, the family, and the child could be conceptualised as a series of concentric circles with the child at the centre. The UNCRC implies that this association should now be understood to be triangular in which the state has a direct responsibility to the child to promote her or his rights. The child has the right to make a direct call on the state and to be heard in the development of legislation and policy, besides receiving protection (2008).

Karin Ringheim similarly affirms that "The responsibilities and duties of parents are to provide direction and guidance in the exercise by the child of his or her human rights" (Ringheim, 2007, 246). Parents have been recast within the very limited role as custodians of their child's best interests, with parental motives vis-à-vis their children mistrusted, and parental decisional authority judged as in need of significant regulatory oversight to ensure that parents act within the stipulated human rights vision of the child's best interests.[4]

The Convention, however, straightforwardly ignored key questions regarding who ought to be appreciated as in authority over children, as well as who is most appropriately situated to define and defend the best interests of the child (e.g., governmental bureaucrats, physicians, healthcare workers, bioethicists, or parents). Parents, for example, must think in terms of the best interests of the family as a whole and, as a result, families accept a wide range of choices that are in the best interests of the family, but not necessarily in the best interests of any particular child (such as, moving to accept a better paying job in a city with greater pollution[5] or an increased crime rate[6]). The goal of the Convention's appeal to basic human rights for children is greatly to restrict the scope of parental authority, forwarding a goal of transforming the traditional family into a cultural and religious historical artifact.[7] Its implications for the bioethics of pediatric decision-making are considerable.

For example, articles 12 and 13 of the Convention assert that children possess freedom of expression in all matters affecting the child, including the "... freedom to seek, receive and impart information and ideas of all kinds, regardless of frontiers, either orally, in writing or in print, in the form of art, or through any other media of the child's choice." As codified in law, such rights seek to defeat the ability of parents actively to censor information and direct their children's education, including sex education. The goal appears to be to limit the ability of traditionally religious parents to raise traditionally religious children. In Europe, claims to such human rights have been successfully used to lower the age of consent for homosexual acts to sixteen-years-of-age. Euan Sutherland in 2000 successfully challenged the British Government in the European Court of Human Rights, leading to a change in the British law (*Sutherland v. The United Kingdom*). In *Gillick v. West Norfold and Wisbeck Area Health Authority* (1985), the British Court held that it was lawful to prescribe contraceptives to young girls under sixteen years-of-age, without the permission or her parents, provided that the child is capable of understanding what is proposed and of expressing her own opinion regarding treatment. The child's liberty rights are perceived as entailing rights to sexual pleasure, generally restricted only by the consent of the parties involved.

The Convention brings into question any religiously directed education of the child and appears opposed to children being raised within a thickly religious or cultural context. Article 14, para 1, asserts the existence of rights to freedom of thought, conscience, and religion.[8] Rather than appreciating the family as the school of moral virtue and religious piety, where parents nurture children in the fundamentals of faith, culture, tradition and duty, the Convention undermines these traditional roles of parents, holding that governments ought only to protect the rights of parents insofar as parents "... provide direction to the child in the exercise of his or her right in a manner consistent with the evolving capacities of the child" (Article 14, para 2). Governments are to ensure that parents guide children towards freedoms of thought, conscience and religion. The Convention treats all religions, cultures, and secular worldviews as each potentially equally good and valuable, positioning minor children (rather than parents) in authority to choose on their own behalf the religion or secular worldview they wish to pursue. This aspect of the Convention bears on

6 Bioethics as Political Ideology

attempts by parents to frame pediatric decision-making on behalf of their children in terms of religious norms.

In addition, children are appreciated as possessing rights to freedom of association and peaceable assembly (Article 15, para 1). No restrictions may be placed on the exercise of such rights, ". . . other than those imposed in conformity with the law and which are necessary in a democratic society in the interests of national security or public safety, public order, the protection of public health or morals or the protection of the freedoms of others" (Article 15, para 2). If a straightforward textual reading of this article is applied to public policy, the Convention appears to be calling for national legal frameworks that prohibit parents from limiting the association rights of their children. This circumstance suggests a fundamental shift in the role of parents, who have usually been appreciated as most appropriately situated to help their children choose friends and associates, while also in authority to limit a child's interactions with others whom the parents judge as inappropriate or dangerous influences on the child. Again, the Convention appears to positions minor children themselves (rather than parents) in authority to make their own judgments about with whom to associate.

Children are held to possess rights to privacy: "No child shall be subjected to arbitrary or unlawful interference with his or her privacy. . ." (Article 16, para 1). Regarding areas of life deemed private, children are often identified as independent of their parents prior to having achieved full maturity. Parental consent to access birth control measures, abortion, education on sexual activities, or testing and treatment for sexually transmitted disease is not appreciated as necessary to protect the best interests of children (Cook et al., 2007). "Sufficiently mature" minors are to be permitted to decide on their own behalf regarding such matters. Here the focus is on the child's "sufficient maturity" rather than on any particular minimum age. Even controversial procedures such as abortion are to be left up to the decision of the child. (For example, current California law requires neither parental consent nor simple notification for a child to obtain an abortion.[9]) Nearly unfettered access to abortion, often provided through state-based taxpayer financing, has been endorsed as central to preserving the human rights of women, as liberating women from the alleged evils of patriarchy and enforced pregnancy.[10] Failing to provide easy access to abortion as a basic entitlement right has been rhetorically recast as a violation of basic human rights.[11] In each case, the minor child is treated in isolation from her parents, thereby undermining the importance of the family and eliminating the assistance of parents and other family members when making difficult and significant medical decisions.

Moreover, member states will ". . . take all effective and appropriate measures with a view to abolishing traditional practices prejudicial to the health of children" (Article 24, para 3). This paragraph has been specifically interpreted to apply to religious circumcision of males, in an attempt to forbid the practice among observant Jews. Although, in a somewhat ironic twist, many international health organizations have begun arguing that male circumcision ought to be encouraged, perhaps made legally mandatory, potentially to reduce the risk of human immunodeficiency virus

(HIV) transmission (Clark et al., 2007; WHO, 2009). It is unclear why secular goals, such as reduction of HIV transmission, would justify routine male circumcision, while religious goals, such as sealing a covenant with God, would not. Provided that the circumcision is performed in a suitably sterile environment with appropriate skill, the surgical action and physical impact on the male child would be identical in either case.

Article 28 provides for a right to a free and compulsory education.[12] Article 29 directs states ideologically to guide such compulsory education so as to reenforce a secular human rights polity:

> States Parties agree that the education of the child shall be directed to: ... (b) The development of respect for human rights and fundamental freedoms, and for the principles enshrined in the Charter of the United Nations; ... (d) The preparation of the child for responsible life in a free society, in the spirit of understanding, peace, tolerance, equality of sexes, and friendship among all peoples, ethnic, national and religious groups...; (e) The development of respect for the natural environment.

Children are to be educated in the virtues of human rights and individual equality as well as to possess a particular regard for the environment. Such compulsory state-based education would appear to be well positioned to lead children away from traditional religions and cultures, towards a universal secular society based on the United Nations' conception of human rights.[13] As Wil Kymlicka observes, the avoidance of such underlying political and moral agendas is precisely why many parents seek exemptions from legal requirements to have their children attend public schools (2001).[14]

In short, the Convention advances a child liberationist agenda: the development of strategies through law and public policy to aid children in forming their own views of morality and human flourishing independently of their parents, to act on their own moral viewpoints in significant areas of life (such as medical decision-making) apart from their parents. Children are appreciated as within parental authority only insofar as parents act as trustees of the child's best interests. Given its suspicion of claims of parental authority over their children, the Convention on the Rights of the Child invites the state to reform parent-child relationships. These commitments lead to a bioethics of pediatric decision-making that marginalizes parental roles.

6.2.2 Example II: Welfare Entitlements to Health

As a second example, consider the United Nations Educational, Scientific and Cultural Organization's (UNESCO) Universal Declaration on Bioethics and Human Rights (issued 19 October 2005),[15] which announced, without argument, a significant array of human rights as welfare entitlements.[16] Article 14 alone announced rights to such purported goods as personal health and healthcare, medicine, nutrition, water, improved living conditions, environmental protection and so forth – as

6 Bioethics as Political Ideology

well as corresponding governmental duties to promote and provide for such public health measures:

1. The promotion of health and social development for their people is a central purpose of governments that all sectors of society share.
2. Taking into account that the enjoyment of the highest attainable standard of health is one of the fundamental rights of every human being without distinction of race, religion, political belief, economic or social condition, progress in science and technology should advance;[17]

   (a) Access to quality health care and essential medicines, especially for the health of women and children, because health is essential to life itself and must be considered to be a social and human good;
   (b) Access to adequate nutrition and water;
   (c) Improvement of living conditions and the environment;
   (d) Elimination of the marginalization and the exclusion of persons on the basis of any grounds;
   (e) Reduction of poverty and illiteracy.[18]

The claim that "... promotion of health and social development for their people is a central purpose of government that all sectors of society share" (paragraph 1), rhetorically places expansive human rights regulatory legislation under the rubric of legitimate political authority, without supposing any need to provide an argument for such legitimacy.

In its "Aims", the Universal Declaration affirms that the intention is "to provide a universal framework of principles and procedures to guide States in the formulation of their legislation, policies or other instruments in the field of bioethics" (p. 76).[19] Section 36 of the Resolutions of the October 2005 General Conference states:

> *The General Conference* ... 1. *Adopts* the Universal Declaration on Bioethics and Human Rights, as annexed hereto; 2. *Calls upon* Member States: (a) to make every effort to adopt measures, whether of a legislative, administrative or other character, to give effect to the principles set out in the Declaration, in accordance with international human rights law; such measures should be supported by action in the sphere of education, training and public information... (p. 74).

The authors' purpose is to introduce new political obligations, statutory rights, welfare entitlements and moral duties under the general rubric of "human rights", through the simple expedient of making the announcement. UNESCO's goal is to delineate a legislative, administrative, and educational agenda for state action.

A central question that must be asked of any governmental regulation, however, is whether it is a legitimate application of moral political authority or an unauthorized act of state bureaucratic coercion.[20] At stake are foundational moral and political questions concerning the limits of governmental authority to interfere within the sphere of the family, to limit the authority of parents, or to intervene in the consensual interaction of persons. There are real limits on legitimate government action to promote goods, including health and healthcare. Governmental establishment of a "human right" to the "highest attainable standard of health" or to "quality health care and essential medicines" constitutes a significant entitlement claim on the goods and services of others. Such regulation seeks coercively to control the conduct of persons to promote the creation and maintenance of such goods, while actively discouraging, if not straightforwardly eliminating, other important goods. It utilizes the time, talent and property of persons to promote the preferred good without their consent and, at times, even against their deeply held moral opposition. There may be significant conflict with the decisions of individuals who do not wish to participate in, or who are sincerely morally opposed to the creation or realization of the politically endorsed good. Imposing such duties on others, including citizens of a state through regulatory activity and taxation, must be justified, non-arbitrary, and demonstrably within the limits of moral political authority.[21] Political announcements of human rights simply do not meet this burden.

6.3 Challenges: Moral, Epistemological, and Political

6.3.1 Moral and Epistemological Ambiguity

The more universally binding moral content, such as the existence of special goods (e.g., freedoms of expression, conscience, religion, health, and so forth), can be established, the easier it is to claim the moral political authority to regulate the conduct of persons to promote the creation and maintenance of such goods. The challenge lies in definitively demonstrating canonical content to ground human rights claims. James Buchanan and Kristen Hessler, for example, attempt to ground human rights in basic human interests:

> We suggest instead that human rights should be understood as moral claims grounded in *basic human interests*. These are interests that are universally shared by all (or nearly all) human beings, and they are the kinds of interests that justify assigning obligations to others, or to society generally, to secure or protect those interests. . . . Health care is a human right because being healthy is a universal interest, common to all people, that grounds duties in others. The basic human interest in health grounds negative duties – duties not to make anyone sick – and, especially for government, positive duties – duties to protect and/or to promote the health of others (2009, p. 213).

There are two fundamental challenges here: first, specifying which needs/interests generate human rights (after all, humans have many needs and interests) and, second, justifying the epistemological and moral jump from (a) the observation that humans have a particular interest/need to (b) the social political conclusion that others (persons or governments) have duties to protect or provide for those interests/needs. They claim that "ultimately, what grounds human rights, including corresponding obligations, is something fundamental to morality: equal consideration for all persons" (2009, p. 213); however, this assertion is insufficient the span the chasm. It only puts off justification or presumes it can be found elsewhere in the defense and specification of the secular fundamentalist claim – the "equal consideration for all persons".

These observations are meant to raise the core epistemological challenge of justifying moral foundations, especially since human rights claims are held to

6 Bioethics as Political Ideology

have significant political implications. Appeal to human rights rhetorically supports the political activism of the Convention on the Rights of the Child, UNESCO's Universal Declaration on Bioethics and Human Rights, and other claims to human rights, without the need to appeal to moral argument, empirical fact, constitutional law or other legal or moral realities. As one advocate on behalf of a supposed human right to a healthy environment asserted: "We have, as individual members of our society, the right, the human right, to a healthy environment. We have this human right because we are a member of our society, our group of people. Our elected representatives have the duty to provide that healthy environment" (Dresler, 2008, p. 701). Similarly,

> As a society, we owe its members the right to a healthy environment, and as individuals we must have the agency to claim this right. . . . it is the duty of the government to provide such human rights. When such a law is passed that provides a healthy environment, then we have incorporated what is considered a human right into a legally protected right. Human rights trump legally protected rights; they are from a higher order. Our government should move to incorporate our human right to a healthy environment by legally protecting us . . . (Dresler, 2008, p. 702).

Or, more generally:

> A global consensus exists on the fundamental importance of human rights. In the sphere of political morality, human rights trump all other considerations, including mandates of law and custom. Nothing can justify violating them (Jaggar, 2009, pp. 156–157).

No argument has been given, nor is any argument expected. Human rights have no real moral or legal foundation, they are simply announced. When a foundation is sought, advocates appeal to yet other pronouncements of human rights, such as the International Convention on Civil and Political Rights, the International Convention on Economic, Social, and Cultural Rights, and the United Nation's Universal Declaration on Human Rights. Similarly, the Convention on the Rights of the Child and the Universal Declaration on Bioethics and Human Rights each just announces itself as building on "human rights" as articulated in the United Nations' Universal Declaration on Human Rights, and other such pronouncements, which have similarly claimed wide-ranging and significant welfare entitlements, moral duties, and governmental powers without argument.[22] In each case, it is a regression from assertion to assertion, to assertion. . .and so forth. Human rights claims appear to have moral, bioethical and political force, in large measure, because of the sheer frequently to which they are referred.

6.3.2 Strategically Ambiguous Appeals to Consensus

Even the routine appeal to a moral "consensus" on human rights is at best strategically ambiguous. Consensus may refer to "unanimity of opinion," "majority of opinion," or "general agreement or accord". Which is required – 100% agreement, two-thirds agreement or a bare majority – honestly to refer to the existence of a moral consensus? Moreover, who constitutes the proper reference class? A moral consensus on a vast array of moral claims can be discovered by limiting the reference class to the proper group of right-thinking experts. In his defense of a basic human right to health care, for example, Buchanan shifts through different types of appeals to the moral authority of "consensus":

1. Consensus as general society-wide moral agreement: "Surely it is plausible to assume that, as with other rights to goods or services, the content of the right must depend upon the resources available in a given society and perhaps also upon a certain consensus of expectations among its members" (Buchanan, 2009, p. 20).
2. Consensus as agreement among a select group of governmental experts: "In the British National Health Service, a consensus on rationing policies develops among medical specialty groups through interaction with regional health authorities, but the public appears to have little opportunity for input and may not even know that rationing is occurring" (Buchanan, 2009, pp. 52–53).
3. Consensus as agreement among "reasonable persons": "an adequate ethical evaluation of an allocation decision may require an assessment of the procedural fairness of the decision-making process (from which that allocation issued) because there is no consensus among reasonable persons as to whether the allocation, considered by itself, is ethically acceptable" (Buchanan, 2009, p. 41).
4. Consensus as agreement among individuals with a similar ideological point of view: "...there is a considerable consensus, at least within liberal democratic political philosophy, that equality of opportunity is a central element of justice" (Buchanan, 2009, p. 62, citing Norman Daniels).
5. Consensus as agreement among a particular group of moral/political theorists: "Although there is also disagreement as to precisely what is included in the notion of procedural justice, there nevertheless is considerable consensus on the core elements ..." (Buchanan, 2009, p. 163).

Each claim to consensus draws on different empirical standards, conceptual meanings, and reference classes of persons for judging the existence of moral consensus; none guarantees connection between the consensus and moral truth. The use of "considerable consensus", in the last example, notes even greater vagueness as a foundation to justify a human right to healthcare. In each case, appeal to "consensus" rhetorically signals moral truth, without the provision of definitive argument. Given disparate reference classes of persons, various background interests and value judgments, divergent accounts of the human right to healthcare (if any) will emerge.

The foundational moral challenge definitively to articulate human rights to health, is that there does not exist any societal consensus regarding the meaning of a right to health or to what constitutes healthcare to which there is a welfare entitlement. A simple literature search is sufficient to illustrate that no such moral consensus regarding basic human rights to health care exists. Orthodox Christians, Orthodox Jews, traditional Moslems, Roman Catholics, and Confucians, simply to site a few examples, do not share a consensus with secular bioethics on human rights, nor do many general secular worldviews. Consequently, members of traditional religions and cultures have good grounds for appreciating the goals and content of the human rights agenda with deep suspicion. It is in direct conflict with their own deeply held beliefs and practices. Human rights education of children seeks directly to undermine the religious and spiritual reality of family life.[23]

6 Bioethics as Political Ideology

For example, insofar as parents act only in the ways sanctioned by the Convention on the Rights of the Child, they will very likely fail to raise children who follow in their religious faith and cultural life. Traditional religions do not permit their children freedom of expression, including sexual expression, religion, thought, and conscience, as if all religious opinions and choices were equally true; nor do they permit their children freedoms of assembly and association. Traditional parents help their children choose their friends, limit their interactions with inappropriate others, and actively censor their television and movie watching, reading material, and internet access. Each decision must be crafted so as properly to orient children toward the good, as their parents understand such good. The family plays a very central role as the core community through which parents teach children how best to live, to aim towards the good, to fulfill their substantial duties to God, family, and others.

The reality of significant moral pluralism and unlimited desires for healthcare, coupled with limited resources – scarcity in healthcare goods and services as well as real financial strain – is causing reassessment worldwide regarding the existence and demands of a right to healthcare. There is widespread disagreement regarding the proper role of government in such matters. Many countries are expanding their private healthcare sectors, permitting individuals to buy out of the increasingly heavily rationed state-based systems (e.g., Germany), while others are placing very significant limits on government financing of healthcare, preferring mandatory family-based healthcare savings accounts (e.g., Singapore). The soaring costs of tax-payer financed welfare systems look ever more economically unsustainable. There is also a growing awareness of the ways welfare programs contribute to the collapse of other important goods, such as the family. Hong Kong, Singapore and China, in particular, have been keen to refashion their healthcare systems so as to affirm traditional family-based medical decision-making.

There is such remarkable moral diversity that it seems implausible that one could find moral consensus, or even two-thirds majority, regarding the existence or content of a welfare entitlement to healthcare. I would not personally bet on a mere preponderance of opinion. Perhaps this explains the turn of some authors to "consensus" not as an indication of moral truth, but as a way to forge a political modus vivendi:

> 6. "I argue, however, that without a politically effective societal consensus on what the right to health care includes . . ." (Buchanan, 2009, p. 145, see also p. 148). Similarly, at the level of international politics: "it requires a collective effort to forge an international consensus on a core set of health entitlements . . ." (p. 229).

Appeal to moral truth in any strict sense has been set aside; rather, a political coalition is being sought adequate to influence the legislative process, to create and enforce a preferred social outcome. Yet, members of such a modus vivendi may possess only limited over-lapping political or economic interests, or share more cynical motives, such as power brokering for professional advantage, dominating the resources, lives and choices of others, or advancing particular social agendas (e.g., sex equality, unfettered access to abortion). It would be a conceptual mistake to infer such acts of political will and vote trading as indications of moral truth.

6.3.3 Rhetorically Shifting the Burden of Proof

These so-called universal declarations on universal human rights encompass an ethical vision that functions on analogy with the universal legislator of Immanuel Kant's categorical imperative or the uniquely privileged and unbiased utilitarian calculator of costs and benefits, who, in either case, purports to derive a canonical understanding of appropriate human choice, rational human preference satisfaction, and legitimate governmental authority from a particular account of moral rationality and rational volition. It is the attempt to authorize and legitimatize state moral authority in terms of a rationally discoverable vision of morality, justice, and proper conduct. This is why it gives accent to claims to special goods, such as "universal human rights" and "health", while also asserting special insight into the human condition. For example, UNESCO's claims, such as "...especially for the health of women and children, because health is essential to life itself and must be considered to be a social and human good", and the Convention's assertions regarding the "best interests" of the child regard what one *ought rationally to understand* as good and morally required, rather than what any particular person, religion, culture, or moral community *actually* considers good or morally required. The Convention on the Rights of the Child, similarly just assumes that *one ought rationally* to appreciate restrictions on the privacy of children, or to their freedoms of religion, association and assembly, to be immoral, and thus that the state should protect such purported rights of children, even against the objections of their parents. Those who disagree are thereby dismissed without argument as immoral and irrational.

Appeal to "fundamental human rights" is an attempt to side-step any actual regional, cultural, community, or religious morality, as well as any legal, customary, or constitutional restrictions on state authority. The underlying assertion is that all humans are morally bound together without either a common political history, confession of faith or shared moral worldview. Consider, for example, David Little's comments on the history of the Universal Declaration on Human Rights:

> After heated debate, the [Human Rights] commission ruled out all confessional references in the Universal Declaration as being inconsistent with the nondiscriminatory character of human rights. Human beings are held to possess human rights, and to be accountable and obliged to live up to them simply because they are human, *not* because they are Muslim, or Christian, or Buddhist, or Jewish, or Hindu, or a member of any particular religious or philosophical tradition. The whole point of human rights is that they are taken to be binding and available, regardless of identity or worldview.
>
> This does not mean, of course, that people are not free to harbor their own personal reasons – religious or otherwise – for believing in human rights. It only means that such views may not be taken as "official" or in any way binding on others who do not share them. Such is the meaning of religious liberty enshrined in the Universal Declaration (2005, p. xi).

The expectation is that through such "universal human rights" secular morality discloses a communality of all persons, justified not in faith, but disclosed through reason itself.

As another commentator put it, while human rights are a Western European Enlightenment concept, the idea of unconditional rights which persons possess simply in virtue of their humanity constitutes "...one of the major achievements of human civilization, much more important than any scientific or technical development" (Andorno, 2007, p. 152). Consequently, when the Convention on the Rights of the Child and the Universal Declaration on Bioethics and Human Rights announce "universal" moral principles and human rights, each thereby announces the secular equivalent of orthodox religious belief. Whereas the "enshrined" meaning of religious liberty permits individuals to believe in human rights for private religious reasons, such religious liberty does not permit individuals and groups NOT to believe in human rights.[24] That particular religious freedom and moral position is forbidden.

The role of human rights language is thus rhetorically to shift the burden of proof away from the governments that coercively impose the social creation of such rights, or tax to maintain such welfare entitlements. Dissenters are thereby made to appear as mean-spirited, idiosyncratic or superstitious, as acting in bad faith, or as fundamentalist adherents of irrational religious dogma. Such a rhetorically shifting of the burden of proof appears to be a core motivation underlying so-called "universal declarations." The point is to control the political and moral agenda while simultaneously claiming legitimacy for governmental regulatory implementation.[25] Indeed, human rights proponents note that among the advantages of utilizing the human rights framework, over and above bioethics and moral analysis, is that it straightforwardly provides a "...legal instrument for implementing policy, accountability, and social justice..." (Reading et al., 2008). Human rights advocates apparently need not be concerned with actual constitutional limitations, or even with having a composed, fair-minded conversation, negotiating with dissenters; instead, they need only utilize the coercive apparatus of the government and the court systems to impose favored solutions.

6.4 The Need for a Canonical Moral Anthropology

What is entirely lacking in the Convention on the Rights of the Child, the Universal Declaration on Bioethics, or other human rights documents, is the articulation of a canonical moral anthropology – the nature and content of the basic goods central to human flourishing, such that one could articulate an account of the "best interests of the child" or of special goods such as "health", without simply begging the question. As a matter of empirical reality, instead of unity one finds a considerable plurality of contradictory and incommensurable, religious and secular accounts of the basic goods central to human flourishing – the goods central to appreciating the "best interests of the child", or for making judgments regarding those "special goods" to which one ought to give accent. One finds as well significantly diverse theories for rationally debating the merits of these divergent understandings of human nature. There appear to be at least as many competing moral anthropologies, with attendant accounts of the basic human goods, as there are major world religions and secular worldviews. Which account of human nature, with whose view of human flourishing and basic goods, should be appreciated as morally normative? One must first specify the normative criteria for determining best interests – that is, how appropriately to balance costs and benefits, rank human goods or cardinal moral concerns. Which consequences ought to be avoided, which virtues inculcated and values embraced, and at what costs? Without such an account, the challenges for knowing how rightly to weigh harms and benefits to determine, for example, the best interests of the child are insurmountable.

Moralizing pronouncements, such as the Convention on the Rights of the Child and the Universal Declaration on Bioethics, are embedded in what can be characterized as a moral cosmopolitanism articulated in the international language of secular modernity; that is, a morality isolated from traditional cultures and religions and devoid of a shared ethics to provide a common moral compass. It a part of a larger project designed systematically to call into question, if not to abandon, traditional religious values and institutions that had previously grounded understandings of moral truth and human flourishing, as well as judgments regarding the best interests of the child.

Such fragmentation of the religious and cultural assumptions that had once framed Western moral judgments, has set the stage for the emergence of deep-seated differences that divide morality and accounts of human flourishing into not merely different, but mutually antagonistic accounts of proper moral conduct. Universal canonical moral judgments cannot be read straightforwardly off of reason, intuition, sense of profanation or moral outrage so as to inform judgments regarding the best interests of the child. As a result, the Convention on the Rights of the Child straightforwardly begs the question in its intent to secure freedoms of expression, speech, religion, conscience, and so forth supposedly to protect the best interests of children. It has no moral grounds or intellectual authority – secular or religious – to support its assertions, much less to sunder children from the authority of their parents.

Similarly, to establish the Universal Declaration's particular understanding of the special good of health and basic human rights to healthcare, and medicine, or the Convention's basic human rights to freedom of speech, conscience, association, religion, and so forth – indeed, to establish any element of the litany of universal moral, bioethical, and political pronouncements – proponents must provide some unequivocal grounding for such universal principles, special goods, and human rights. What are their origins, foundations, or limits? Is it an appeal to personal preferences or moral intuitions, to current biomedical convention, to consequences, casuistry, the notion of unbiased choice, game theory, or middle-level principles? Are they somehow inherent in human nature? All such attempts, as H.T. Engelhardt Jr. argues (1996), confront insurmountable obstacles: one must already presuppose a particular morality so as to choose among intuitions, rank consequences, evaluate exemplary cases, define an account of human nature, or mediate among various principles, otherwise one will be unable to make any rational choice at all. As he points out, even if one merely ranks cardinal moral concerns, such as liberty, equality, justice and security – or health, for that matter – differently one affirms different moral visions,

6 Bioethics as Political Ideology   113

divergent understandings of the good life, varying senses of what it is to act appropriately (1996, pp. 40ff). To ignore such on-going and fundamental differences of moral viewpoint, while declaring a "universal bioethics" and "human rights" sufficient to establish the moral political authority to impose this particular moral vision just assumes what must be proven. Once secularized and thereby cut off from any particular metaphysical anchor (e.g., God), bioethics and human rights are no more and no less than what humans make of them.

It is simply assumed that moral authority exists to utilize the time, talent, and property of others, to tax and regulate, to promote the preferred welfare entitlements.[26] As Engelhardt argues, bioethics continues to give political expression to the Enlightenment hope that ethics, and thus bioethics, ". . .should liberate from unjustified customs, and constraints, those contrary to the demands of universal moral reason" binding all persons together in a common secular morality (2006, p. 20). As a result, as a field bioethics inevitably seeks universal declarations on a wide variety of medical moral controversies, such as abortion, embryonic stem cell research, cloning, genetic enhancement, experimentation on human embryos, the place of religious thought in ethical analysis, and the sale of human organs for transplantation (Cherry, 2005). In this fashion, bioethicists and human rights advocates have been collaborating in the attempt to elaborate an international bioethics to guide court decisions, public deliberation, clinical decision making, and legislative action, as well as international convention and treaty. Substantial moral disagreement is to be shunned, if not actively persecuted.[27] In summary, bioethics and human rights routinely functions at the level of conceptual ideology: "a prescriptive doctrine that is not supported by rational argument"; each is a Nietzschean will to power, asserting moral political authority to impose its views without real legitimacy.

6.5 Conclusion

If the field of bioethics is to extricate itself from the culture wars, as Fox and Swazey admonish, then bioethics must seek an authoritative account of moral political authority that does not simply assume what needs to be proved. Even coercive taxation to support a particular vision of distributive justice and welfare entitlements, or to secure the human rights of the child, requires legitimate moral political authority. As noted, it cannot appeal to a universal, content-full moral understanding of the special good of health, or pretend to secure justification for political authority to impose such a vision though so-called basic human rights, without begging the question.

Here free-markets can be appreciated as drawing moral authority not from assertions of so-called moral consensus, ideal theories of rational action, or even deep moral intuitions regarding consequences, human rights and dignity, or other cardinal moral concerns, but from the agreement of the parties themselves to collaborate. Collaborators need not agree regarding the background ranking of values or moral principles, cultural or religious assumptions; they need only affirm the content of their agreement. No value standard or moral vision must be presumed, just the recognition that collaboration is possible through agreement. Agreement is the ground of the moral justification of such collaboration. In moral theoretical terms, agreement or permission can be appreciated as a deontological moral concern, focused on a right making condition that governs independently of a concern to realize a particular good or to secure a positive balance of benefits over harms.

Given the background of competing moral visions, accounts of human rights, and the nature and content of special goods, free-markets are central for understanding authoritative human interaction. Market transactions and contractual relationships draw moral authority from the consent of the participants to be bound by their agreement. The parties to the transactions freely convey authority to the enforcement of the specified conditions. Moral authority to interfere in the interaction of consenting persons is created by, and thus limited to, the actual agreements of actual persons. Markets are not affirmed as good in themselves, nor are they affirmed as good because they lead to greater innovation or otherwise maximize the good. Instead, markets create social space for unencumbered authoritative human interaction. In the absence of the ability definitively to demonstrate a particular moral content as the sole legitimate basis for moral political authority, without begging the question, markets are simply the result of respecting the moral authority of persons over themselves and their private property.

Secular morally authorized governmental statutes are confined to the protection of persons' rights to forbearance, protection of individuals from battery, and to such additional policies to which actual persons give actual consent. Persons, and organizations of persons, may be held responsible for nonconsensual acts of violence, fraud, and breach of contract, since such actions violate the rights of persons not to be touched or used without permission. However, the rights of persons over themselves, and even over their children, will foreclose what many envision to be worthwhile goals. Free-markets appreciate persons as only entitled to their own private resources and to those additional resources that they are able peaceably to convince others to donate or sell to them. Basic human rights to health or health care do not exist; nor do non-emancipated children possess rights to freedom of expression, conscience, religion, association, and speech over against their parents.

Freely chosen, market-based health care, including financing, development, procurement, and distribution of pharmaceuticals, medical devices, goods and services, respects the liberty and integrity of persons to pursue their own deep moral commitments. Markets create social and political space for the free interaction of individuals. They secure the possibility of real moral diversity within particular states as well as for the emergence of a worldwide network of non-geographically based communities with their own particular understandings of moral probity, including bioethics, health care policy and institutional restrictions. Substantial moral, ideological, and religious diversity will likely exist not owing to any endorsement of diversity's value, but because legitimate moral authority does not exist to foreclose peaceable consensual interaction. Coercively taxing others to create a comprehensive state-based system of welfare entitlements and health care rights

6 Bioethics as Political Ideology

is prohibited, not because of the simple ascription of positive individual rights, but straightforwardly because of the lack of moral authority to so tax (Bole, 2004). The permission of the parties in the marketplace to collaborate creates, rather than discovers, moral authority.

If bioethics is to capture a "...composed, fair-minded, rationally thoughtful outlook ideally associated with a philosophically oriented field" (Fox and Swazey, 2005, p. 369), it must think itself out of the box of its political ideology of human rights, welfare entitlements, and top-down governmental regulation driven towards a fully secular vision of ethics and social justice. Instead of presuming to see itself as in authority to guide public policy, medical decision making, or international human rights law, to give testimony in courts of law, or coercively to force parents to raise their children in line with current secular fashionable ideals of political correctness, bioethics must re-imagine itself as the market place of ideas, including moral and religious ideas, where conceptual and moral analyzes are critically and carefully explored and real moral diversity is respected and defended, rather than simply denied and legislatively dismissed through the simple expedient of rhetorical appeals to "human rights" and "global consensus". It must cease to be an ideologically driven will to power, and recapture the social space necessary for both robust expression and instantiation of diverse moral, religious, and secular worldviews. That Fox and Swazey seek a closer association between the field of bioethics and international human rights law is not remarkable; what is astonishing is that they believe that this will somehow extricate bioethics from the culture wars. International human rights law and bioethics are frequently deeply interwoven and exist at the very heart of the culture wars.

Notes

1. Fox and Swazey cite an unpublished paper of the liberal bioethicist Daniel Callahan in their support: "It will have to remember that it is possible for bioethicists to be wrong without being immoral, that bioethics harms itself if it turns into a moral crusade, either for the values of the left or the right, and that a healthy bioethics will be one where conservatives and liberals understand they have a common cause, one best pursued in lively dialogue rather than as opposing armies" (Callahan, 2003; see also 1997); cited in Fox and Swazey, 2005, p. 369.
2. The term "culture wars" was coined by James Davidson Hunter in *Culture Wars: The Struggle to Define America* (1991).
3. The American Society for Bioethics and Humanities, for example, seeks both to specify the core competencies of bioethics consultation and to certify bioethics consultants, precisely so that it can institutionalize a particular secular vision of bioethics and social justice while also regulating access to jobs in the field. For its account of the core competencies and approach for clinical ethics consultation see ASBH, 1998; for analysis in favor of having ASBH certify bioethics consultants see Tarzian (2009), Kipnis (2009), and Spike (2009); for critical arguments against see Bishop et al. (2009), Childs (2009) and Engelhardt (2009).
4. The "best interests of the child" are to be judged in terms of the child's "human rights", rather than in terms of what parents consider best for their children: "The responsibilities and duties of parents are to provide direction and guidance in the exercise by the child of his or her human rights" (Ringheim, 2007, p. 246).

5. Airborne particulate pollution reportedly plays an important role in human health. According to one study of six small cities, residents of more polluted cities have increased risk of premature morbidity, as compared to other less polluted cites. Residents of Steubenville, Ohio, the most polluted city studied, had as much as a 26% increased risk of premature death, compared to the cleanest cities studied (Dockery et al., 1993). Investigators found an association between increased risks and differences in ambient fine particle concentrations of 18.6 micrograms per cubic meter of air between the most and least polluted cities. This study focused solely on particulate air pollution, adjusting the study design to account for other personal exposures, such as occupational contact and smoking habits. A California research team concluded that children who live within about a third of a mile of busy freeway traffic have decreased lung function and lung growth (Gauderman, 2007; Sandström and Brunekreef, 2007). The National Morbidity, Mortality, and Air Pollution Study (NMMAPS) argued that in large American cities increases in particulate pollution led to increased hospital admissions for cardiovascular disease, pneumonia and chronic pulmonary disease. The study examined five of the most widespread outdoor air pollutants: ozone, sulfur dioxide, nitrogen dioxide, particulate matter, and carbon monoxide (Samet et al., 2000).

6. After the devastation caused by hurricane Katrina, many families moved back to New Orleans, for example, despite the fact that it has been ranked by *City Crime Rankings 2008–2009* as having the highest per capita crime rate with 209 murders, and a 199.1% increase in violent crime during the past year (O'Leary, Morgan, and Boba, 2008).

7. That the goal of the Convention and its proponents is to liberate children from their parents is clear: "Debates surrounding the rights of adolescents to receive confidential and private reproductive health services have centered around the potentially conflicting interests of parents and their children. The desire of parents to guide and direct their children's health and development and make health-care decisions for their children is easily understandable. However, the health threat faced by adolescents expose the tension between public or societal interests in maintaining a health population and private or parental interests in maintaining control over their children" (Ringheim, 2007, p. 245).

8. "States Parties shall respect the right of the child to freedom of thought, conscience and religion" (Article 14, para 1).

9. "Do I have to get my parent's permission to get an abortion? No. You do not need anyone's permission, and the law protects your privacy. No one else has the right to know or do anything about it – not your parents, your boyfriend or partner, or your husband. Even if you are married or under 18, the decision is up to you." ACLU, "Your Health; Your Rights" [On-line] Available: www.teensource.org.

   Jutta M. Joachim documents the real anger directed at the Vatican for blocking various proposals at the International Conference on Population and Development in Cairo, 1994, that focused on easy access to abortion and contraception. "To prevent the inclusion of reproductive rights and reproductive health on the conference agenda, the Holy See together with Catholic countries responded by entering into an alliance with fundamentalist Islamic countries, such as Iran, Libya, and the Sudan, despite their ideological differences. The alliance successfully blocked discussion on Chapters 7 and 8, devoted to women's reproductive rights and health, by making ample use of UN rules and procedures, such as frequent oral interventions and the demand for brackets, indicating their disagreement. . . .While the opposition of the Holy See was a significant obstacle, it remains unclear why the Women's Alliance was unable to overcome it" (2007, pp. 150–151).

10. The United Nations guideline, "HIV/AIDS and Human Rights" (1996), specifically calls for easy access to abortion: "Laws should also be enacted to ensure women's reproductive and secular rights, including the right of . . . means of contraception, including safe and legal abortion and the freedom to choose among these, the right to determine the number and spacing of children . . ."

11. Here one might also consider the *Protocol to the African Charter on Human and People's Rights on the Rights of Women in Africa*: ". . .protect the reproductive rights of women by

6 Bioethics as Political Ideology

authorizing medical abortion in cases of sexual assault, rape, incest, and where the continued pregnancy endangers the mental and physical health of the mother or the life of the mother or the foetus" (African Union, 2003, article 14 c). Failure to provide easy access to legal abortion has even been rhetorically recast as a violation of women's "right to life": "Securing abortion rights is indispensable to guaranteeing women's right to life...Nearly 600,000 women die each year of complications related to pregnancy and birth. ... even when legal abortion is available for pregnancies that are directly life-threatening, women continue to seek illegal abortions from which they suffer injury and death. Thirteen to twenty percent of the 600,000 annual deaths related to pregnancy and birth result from unsafe abortions and most of these lives could have been preserved if safe abortions had been legally available" (Jaggar, 2009, p. 144). The author leaves conveniently unstated any discussion of the moral responsibility of women for having choosen to engage in an illegal black market practice. Having already judged abortion to be good, Jagger proceeds to condemn any restriction on easy assess to the practice.

12. "States Parties recognize the right of the child to education, and with a view to achieving this right progressively and on the basis of equal opportunity, they shall in particular: (a) Make primary education compulsory and available free to all; (b) Encourage the development of different forms of secondary education, including general and vocational education, make them available and accessible to every child and take appropriate measures such as the introduction of free education and offering financial assistance in case of need. . ." (Article 28).

13. Consider the Minnesota Human Rights Education project: "This is my home": [On-line] Available: www.hrusa.org/thisismyhome/. This is a joint program of the Minnesota Department of Human Rights and the University of Minnesota Human Rights Resource Center designed to integrate ideologically oriented human rights education into the kindergarten through high-school curriculum statewide, so as systematically to reeducate children away from the traditional moral, religious, and cultural assumptions of their parents. For example, it teaches that state action, such as affirmative action, is necessary to redress supposed past wrongs, such as discrimination based on sexual orientation: "Affirmative Action – Action taken by a government or private institution to make up for past discrimination in education, work, or promotion on the basis of age, birth, color, creed, disability, ethnic origin, familial status, gender, language, marital status, political or other opinion, public assistance, race, religion or belief, sex, or sexual orientation".

14. The aim is "to integrate citizens into a modern societal culture, with its common academic, economic, and political institutions, and this is precisely what ethno-religious sects wish to avoid. Moreover, the sorts of laws from which these groups seek exemption are precisely the sorts of law which lie at the heart of modern nation-building – for example, mass education (Kymlicka, 2001, p. 37).

15. UNESCO joined a growing number of national and international organizations producing so-called "universal" statements on bioethics and healthcare policy. See, for example, World Medical Association (1975, 1983, 1989, 1996, 2000, 2002, 2004), Council of Europe (1997), American Society for Bioethics and Humanities (1998), and World Health Organization (1991).

16. The Universal Declaration on Bioethics and Human Rights sets out without argument, defense, or detailed analysis a series of "bioethical principles" including, among numerous others:

(1) quasi-forbearance rights: individual-oriented informed consent (article 6), and respect for privacy and confidentiality (article 9);

(2) positive claim rights: such as, respect for autonomy (article 5), human dignity and other (unspecified) human rights (article 3.1), equality, justice, and equity (article 10), solidarity and cooperation (article 13), the promotion of health, access to adequate food and water, improving living conditions, reducing poverty and illiteracy (article 14), access to health care and scientific knowledge (article 14),

(3) other-regarding duties: such as placing the priority of the interests and welfare of the individual over the sole interest of science or society (article 3.2); working to benefit while minimizing harms to research participants (article 4); paternalistically protecting those unable to consent (article 7), non-discrimination and non-stigmatization (article 11); as well as respect for personal integrity (article 8), those deemed vulnerable (article 8), cultural diversity and pluralism (article 12), protecting future generations (article 16) and the environment (article 17).

(4) institutional ethical principles: the creation of ethics committees (article 19) and opportunity for informed pluralistic public debate (article 18).

The principles are for the most part undefined. The "right to health" being one of the few exceptions. According to one member of the UNESCO committee that drafted the "Universal Declaration", such ambiguity was intentional:

> This method, which may surprise some, is in fact a common practice in law, in conformity with the old maxim "Omnis definitio in jure periculosa est". (Every definition in law is perilous.) ... In the case of the UNESCO Declaration, this strategy can also be explained for practical reasons, because it would have been impossible to reach a global agreement on the precise meaning of terms such as "human dignity", "autonomy", "justice", "benefit", "harm" or "solidarity", which have a long philosophical history and are, to some extent, conditioned by cultural factors. Thus, the generality in the formulation of the principles can ultimately be justified by the need to find a balance between the universalism of some bioethical norms and the respect for cultural diversity (Andorno, 2007, p. 150).

17. Here UNESCO is implicitly drawing on the popular idea that health as a fundamental human right is essential to equal worth of individual liberty. Wide ranging welfare rights to adequate nutrition, shelter, education, and health care are underscored as basic goods necessary successfully to secure individual abilities and choices. Because health care helps guarantee the chance to enjoy the normal range of human opportunities, access to health care as a basic human right, without race, religion, economic or national barriers, is given special status as foundational to preserving liberty and equality of opportunity. Here, the state is envisaged as a liberal morally pluralistic democracy, framing economic, institutional, and social structures that guarantee equal access to the basic goods of society (healthcare, education, and other welfare entitlements), so as to sustain fair equality of opportunity and outcome. Market-based systems are decried as advantaging the healthier and wealthier segments of society (see, e.g., Daniels, 2007). For a positive account of the free market in medicine and the development of medical technologies see Cherry (2003) and Meadowcroft (2003).

18. It is worth noting the similarities between the assertions of the Universal Declaration on Bioethics and Human Rights and other statements from the United Nations, such as the Commission on Human Rights, which similarly stated: The "right of everyone to the enjoyment of the highest attainable standard of physical and mental health is a human right" (United Nations Office of the High Commissioner for Human Rights, The right of everyone to the enjoyment of the highest attainable standard of physical and mental health, 2004).

19. Indeed, many argue that the Universal Declaration on Bioethics and Human Rights did not go far enough in setting out an obligatory agenda for presumably independent states: "The draft Declaration produced by the IBC had many shortcomings, but at least it tried to advance international bioethics by introducing new obligations for states, such as sharing of benefits and transparency of decision making. The latest draft purposely eliminates all new obligations of states, replacing "shall" with "should" everywhere except where states are already bound by international agreements or where actions are required of UNESCO itself" (Williams, 2005, p. 211).

20. As John Locke pointed out, governments that use the property and persons of their citizens without permission are conceptually no different than thieves: "Wherever law ends, tyranny begins, if the law be transgressed to another's harm; and whosoever in authority exceeds

6 Bioethics as Political Ideology

the power given him by the law, and makes use of the force he has under his command to compass that upon the subject which the laws allows not, ceases in that to be a magistrate, and acting without authority may be opposed, as any other man who by force invades the right of another" (1980 [1690], p. 103 §202).

21. Rights, even rights that are proclaimed universal fundamental human rights, do not and cannot exist in a social and economic vacuum. Rights place burdens on others, often without their consent. As Richard Epstein makes the point, "To create a right in A is to impose a correlative duty on some other person. That person may be specifically identified, as when A has the right to collect $10 from B. But all too often, the identity of the person who bears the correlative duty is concealed. The most common tactic used to achieve that false separation is to interpose some impersonal entity between the holder of the right and the bearer of the correlative duty. Thus the correlative duty to A's right could fall on some corporation or on the government" (1997, pp. 5–6). But, as Epstein notes, placing a duty on corporations and governments is just an opaque way of placing a duty on persons: shareholders and citizens. In short, governmental regulation in the name of a right to the "highest attainable standard of health" constitutes a significant claim to the goods and service of others.

22. The Universal Declaration on Bioethics and Human Rights states: "*Recalling* the Universal Declaration of Human Rights of 10 December 1948, the Universal Declaration on the Human Genome and Human Rights adopted by the General Conference of UNESCO on 11 November 1997 and the International Declaration on Human Genetic Data adopted by the General Conference of UNESCO on 16 October 2003" (UNESCO, 2005, p. 75).

23. For an additional example, consider the ways in which human rights advocates seek fundamentally to revolutionize background traditional understandings of the place, meaning, and reality of sexuality and marriage Proponents of same-gender marriage, for example, assert the existence of basic human rights of homosexual couples to state recognized and socially respected "marriages". See Human Rights Campaign Urges LGBT Community Activity: Continue the Momentum, Lobby Incoming Members of the 111th Congress. January 6, 2009. [On-line]. Available: www.hrc.org/11831.htm. James Davidson Hunter reflects on why homosexual marriage is such a flash point of the culture wars: "The answer goes right to the heart of assumptions about the moral order: what is good, what is right, what is appropriate. Family life, however, is also a "school of virtue," for it bears the responsibility, as no other institution can, for socializing children – raising them as decent and moral people, passing on the morals of a community to the next generation. How parents view nature in matters of sexuality, therefore, is reflected in the ways they teach children about right and wrong. How the actors in the contemporary culture war view nature in matters of sexuality, in turn, will be reflected in their different ideals of how the moral order of a society will take shape in the future" (1991, pp. 188–189). If marriage is appreciated as essentially a contractual relationship among consenting persons, it becomes plausible to recast marriage as primarily a legal arrangement regarding the rights and duties of community living, community property, health insurance, and inheritance rights. Here, marriage is no longer appreciated as holy matrimony, aiming at deep spiritual and religious goods, but has been relocated as a legal and civic arrangement. The goods of marriage have been removed from their religious and spiritual context, and have been reduced through civic ritual to the immanent experiential goods of this world. In general secular terms, there is no reason to think of marriage as other than a living arrangement among consenting persons. Persons who live within the thick moral communities of traditional religions and cultures will recognize such a shift as an immanent displacement of the transcendent: marriage has been recast as a civic or legal arrangement aimed at goods in this world, rather than a spiritual unity aimed at salvation.

24. For accounts of the role of religion in public health care debate which do not rely on the rhetoric of secular "basic human rights" see Cherry (2007), Tollefsen (2007), Beckwith (2007).

25. That such underlying motivation exists is clear. Consider the relationship between non-governmental organizations and the United Nations general assembly: "...[T]he General

Assembly's agenda has a "collective legitimization function." It defines what actions governments ought to engage in and what actions they ought to abstain from in the international arena. Its collective legitimization is also the reason why NGOs consider the agenda to be an appealing target. The backing of the international community gives weight to the demands of these otherwise weak actors at the domestic level" (Joachim, 2007, p. 17).

26. Consider also the politics of abortion. As noted above, rights to privacy have been successfully utilized to defend the "rights" of children to obtain abortions without their parents consent. Similarly, in *Planned Parenthood v. Casey* [1992]), the United States Supreme court argued that in *Roe v. Wade* (1973) the significance of individual control over the self and to personal bodily integrity had been correctly applied to a woman's right to choose abortion. Concerns to preserve the unborn living fetus, even in cases of late-term partial birth abortion, failed to satisfy the burden of proof, according to the court, to override the significance of individual authority over one's body. Respect for individual freedom, the court held, is of greater significance than human life. Nearly unfettered access to abortion, often through state-based taxpayer financing, has been endorsed as central to preserving human rights, as liberating women from the alleged evils of patriarchy and enforced pregnancy. The United Nations guideline, "HIV/AIDS and Human Rights" (1996), specifically calls for easy access to abortion: "Laws should also be enacted to ensure women's reproductive and secular rights, including the right of . . .means of contraception, including safe and legal abortion and the freedom to choose among these, the right to determine the number and spacing of children. . ." Failing to provide easy access to abortion as a basic entitlement right is again rhetorically recast as a violation of basic human rights.

27. See Messikomer et al. (2001) for a brief discussion of the ways in the contemporary bioethics seeks to screen out religious bioethics dialogue in favor of a search for consensus within general secular morality.

References

___. 2008. Ideology. *Oxford English Dictionary*, On-line edition. Accessed 10 August 2011.

African Union. 2003. *Protocol to the African charter on human and people's rights on the rights of women in Africa*. Maputo: African Union.

American Society for Bioethics and Humanities. 1998. *Core competencies for health care ethics consultation*. Glenview, IL: American Society for Bioethics and Humanities.

Andorno, R. 2007. Global bioethics at UNESCO: In defence of the Universal Declaration on Bioethics and Human Rights. *Journal of Medical Ethics* 33: 150–154.

Beckwith, F.J. 2007. Bioethics, the Christian citizen, and the pluralist game. *Christian Bioethics* 13: 159–170.

Bishop, J., J.B. Fanning, and M.J. Bliton. 2009. Of goals and goods and floundering about: A dissensus report on clinical ethics consultation. *HEC Forum* 21(3): 275–291.

Bole, T.J., III. 2004. The perversity of Thomisitc natural law theory. In *Natural law and the possibility of a global ethics*, ed. M.J. Cherry, 141–147. Dordrecht: Kluwer.

Buchanan, A. 2009. *Justice & healthcare: Selected essays*. New York: Oxford University Press.

Callahan, D. 1997. Bioethics and the culture wars. *The Nation* 264(14): 23–24.

Callahan, D. 2003. Bioethics and the American culture wars. (Unpublished paper).

Cherry, M.J. 2003. Scientific excellence, professional virtue, and the profit motive: The market and health care reform. *The Journal of Medicine and Philosophy* 28(3): 259–280.

Cherry, M.J. 2005. *Kidney for sale by owner: Human organs, transplantation, and the market*. Washington, DC: Georgetown University Press.

Cherry, M.J. 2007. Traditional Christian norms and the shaping of public moral life: how should Chrisitans engage in bioethical debate within the public forum? *Christian Bioethics* 13(2): 129–138.

6 Bioethics as Political Ideology

Childs, B.H. 2009. Credentialing clinical ethics consultants: Lessons to be learned. *HEC Forum* 21(3): 231–240.

Clark, P.A., J. Eisenmann, and S. Szapor. 2007. Mandatory neonatal male circumcision in Sub-Saharan Africa: Medical and ethical analysis. *Medical Science Monitor* 13(12): RA205–13.

Cook, R.J., J.N. Erdman, and B.M. Dickens. 2007. Respecting adolescents' confidentiality and reproductive and sexual choices. *International Journal of Gynaecology and Obstetrics* 98: 182–187.

Council of Europe. 1997. Convention for the protection of human rights and dignity of the human being with regard to the application of biology and medicine: convention on human rights and biomedicine Oviedo. [On-line] Available: http://conventions.coe.int/treaty/en/treaties/html/164.htm. Accessed 5 July 2007.

Daniels, N. 2007. *Just health: Meeting health needs fairly*. Cambridge: Cambridge University Press.

Dockery, D.W., C.A. Pope, X. Xu, J.D. Spengler, J.H. Ware, M.E. Fay, B.G. Ferris, and F.E. Speizer. 1993. An association between air pollution and mortality in six U.S. cities. *New England Journal of Medicine* 329: 1753–1759.

Dresler, C. 2008. A clash of rights: should smoking tobacco products in public places be legally banned? *Annals of Thoracic Surgery* 86: 700–702.

Engelhardt, H.T., Jr. 1996. *The foundations of bioethics, second edition*. New York: Oxford University Press.

Engelhardt, H.T., Jr. 2006. The search for a global morality: Bioethics, the culture wars, and moral diversity. In *Global bioethics: The collapse of consensus*, ed. H.T. Engelhardt, Jr. Salem, 18–49. MA: M&M Scrivener Press.

Engelhardt, H.T., Jr. 2009. Credentialing strategically ambiguous and heterogeneous social skills: The emperor without clothes. *HEC Forum* 21(3): 293–306.

Epstein, R. 1997. *Mortal peril: Our inalienable right to health care?* New York: Addison-Wesley.

Fox, R.C., and J.P. Swazey. 2005. Examining American bioethics: Its problems and prospects.*Cambridge Quarterly of Healthcare Ethics* 14: 361–373.

Gauderman, W.J. 2007. Effect of exposure to traffic on lung development from 10 to 18 years of age: A cohort study. *Lancet* 369(9561): 571–577.

Hunter, J.D. 1991. *Culture wars: The struggle to define America*. New York: Basic Books.

Jaggar, A.M. 2009. Abortion rights and gender justice. In *Abortion: Three perspectives*, eds. M. Tooley, C. Wolf-Devine, P.E. Devine, and A.M. Jaggar, 120–175. New York: Oxford University Press.

Joachim, J.M. 2007. *Agenda setting, the UN, and NGOs: Gender violence and reproductive rights*. Washington, DC: Georgetown University Press.

Kipnis, K. 2009. The certified clinical ethics consultant. *HEC Forum* 21(3): 249–261.

Kymlicka, W. 2001. Western political theory and ethnic relations in Eastern Europe. In *Can liberal pluralism be exported? Western political theory and ethnic relations in Eastern Europe*, eds. W. Kymlicka and M. Opalski, 13–106. New York: Oxford University Press.

Little, D. 2005. Foreword. In *For all peoples and all nations: The ecumenical church and human rights*, ed. J.S. Nurser, ix–xii. Washington, DC: Georgetown University Press.

Locke, J. 1980[1690]. *Second treatise of government,* ed. C.B. Macpherson. Indianapolis, IN: Hackett.

Meadowcroft, J. 2003. The British National Health Service: Lessons from the 'socialist calculation debate'. *The Journal of Medicine and Philosophy* 28: 307–326.

Messikomer, C.M., R.C. Rox, and J.P. Swazey 2001. The presence and influence of religion in American bioethics. *Perspectives in Biology and Medicine* 44(4): 485–508.

O'Leary Morgan, K., S. Morgan, and R. Boba. 2008. *City crime rankings 2008–2009: Crime in metropolitan America*. Washington, DC: CQ Press.

Planned Parenthood v. Casey. 1992. 505 US 833, 857–859.

Reading, R., S. Bissell, J. Goldhagen, J. Harwin, J. Masson, S. Moynihan, N. Parton, P. Santos, M. Thoburn, and E. Webb. 2008. Promotion of children's rights and prevention of child maltreatment. *The Lancet*, e-publish ahead of print, December 3, 2008.

Ringheim, K. 2007. 'Ethical and human rights perspectives on providers' obligation to ensure adolescents' rights to privacy. *Studies in Family Planning* 38(4): 245–232.

Roe v. Wade. 1973. 410 US 113.

Samet, J.M., F. Domnici, F.C. Curriero, I. Coursac, and S.L. Zeger. 2000. Fine particulate air pollution and mortality in 20 U.S. cities, 1987–1994. *New England Journal of Medicine* 343(24): 1742–1749.

Sandström, T., and B. Brunekreef. 2007. Traffic-related pollution and lung development in children, *Lancet* 369(9561): 535–537.

Spike, Jeffrey, P. 2009. Resolving the vexing question of credentialing: Finding the Aristotelian mean. *HEC Forum* 21(3): 263–273.

Sutherland v. The United Kingdom. 2001. *Human Rights Case Digest* 12(3–4): 199–200.

Tarzian, A.J. 2009. Credentials for clinical ethics consultation – are we there yet? *HEC Forum* 21(3): 241–248.

Tollefsen, C. 2007. Religious reasons and public healthcare deliberations. *Christian Bioethics* 13(2): 139–158.

United Nations. 1996. *HIV/AIDS and human rights*. New York: United Nations.

United Nations Educational, Scientific and Cultural Organization (UNESCO). 2005. Universal declaration on bioethics and human rights. In *Records of the general conference,* 74–80 Geneva: UNESCO. [On-line]. Available: http://portal.unesco.org/en/ev.php-URL_ID=31058&URL_DO=DO_TOPIC&URL_SECTION=201.html. Accessed 5 Jul 2007.

United Nations Office of the High Commissioner for Human Rights. 1990. *Convention on the rights of the child*. New York: United Nations.

United Nations Office of the High Commissioner for Human Rights 2004. *The right of everyone to the enjoyment of the highest attainable standard of physical and mental health. United Nations Commission on Human Rights Resolution 2004/27*. New York: United Nations.

Williams, John R. 2005. UNESCO's proposed declaration on bioethics and human rights – A bland compromise. *Developing world bioethics* 5(3): 210–215.

World Health Organization. 1991. Human organ transplantation: A report on the developments under the auspices of the WHO. *International Digest of Health Legislation* 42: 389–396.

World Health Organization. 2009. Operational guidance for scaling up male circumcision services for HIV prevention. Geneva, Switzerland: WHO. Available: www.who.int/hiv/pub/malecircumcision/op_guidance/en/index/html. Accessed 22 July 2009.

World Medical Association. 1975, 1983, 1989, 1996, 2000, 2002, and 2004. *Declaration of Helsinki*. [On-line] Available: www.wma.net/e/policy/b3.htm. Accessed 5 July 2007.

Chapter 7
The "s" in Bioethics: Past, Present and Future

Ana S. Iltis and Adrienne Carpenter

What is bioethics? Who is a bioethicist? Is bioethics an area of inquiry, a discipline, or a field? Are bioethicists members of a distinct profession? Some see bioethics as an area where individuals from different professions "meet" to inquire about ethical issues associated with science, medicine, technology, and health care. Some see bioethicists as professionals who share goals, and have specific socially recognized roles and functions for which they ought to be certified, accredited or at least bound by a code of ethics. Others hold that we ought to work toward such a state in which bioethicists are professionals with a defined and robust ethos, shared goals and perhaps even advocate for particular moral positions together through their professional organization. Each of these views proposes a particular meaning for bioethics and its practitioners. In this chapter, we first argue that such views represent a desire for bioethics to mean something very particular. Second, we argue that bioethics cannot live up to this expectation, given its history, its functional and disciplinary diversity, and so forth. In short, we demonstrate that there are multiple ways in which bioethics is plural or diverse. Each of these has implications for bioethics education and for the life of an organization such as the American Society for Bioethics and Humanities (ASBH).[1] We argue that any general discussion of what it means to educate bioethicists would be comparable to having a conversation about educating teachers without specifying whether we are educating teachers who will teach particular age groups or subjects, children with special needs, and so forth. Similarly, to talk about the appropriate goals and scope of a large bioethics organization would be like discussing an organization that includes anyone who is a teacher, or even anyone who has an interest in education, e.g., public school teachers, Montessori teachers, Waldorf teachers, homeschooling parents, dance teachers, scholars who study educational systems or pedagogy and so on. Third, we argue that this diversity is to be respected and protected. The "s" in bioethics is significant not merely because it describes the current state of the enterprise but it describes an appropriate future for the enterprise.[2]

A.S. Iltis (✉)
Center for Bioethics, Health and Society, Wake Forest University, Winston-Salem, NC 27109, USA
e-mail: iltisas@wfu.edu

H.T. Engelhardt, Jr. (ed.), *Bioethics Critically Reconsidered*,
Philosophy and Medicine 100, DOI 10.1007/978-94-007-2244-6_7,
© Springer Science+Business Media B.V. 2012

The drive to secure a particular robust account of bioethics that admits of less diversity than that which reality delivers may be understood best in light of the socio-political and socio-economic contexts in which bioethics has operated. Bioethics has become not only an area of inquiry but an activity resulting in recommendations at the individual, institutional, state, federal and international levels regarding, for example:

- when to declare a person dead (see, for example, Boucek et al.) where "on the basis of the recommendations of the ethics committee" would-be donors were declared dead 1.25 minutes after cardiocirculatory function so that organ procurement procedures could begin earlier (2008, p. 711)
- who should make medical decisions when a patient is incapacitated and how such decisions should be made (see, for example, Buchanan and Brock, 1990)
- what mechanisms are and are not licit for increasing the number of organs available for transplantation (see, for example, IOM, 2006)
- the permissibility of using human embryos for stem cell research or other endeavors (see, for example, National Bioethics Advisory Commission, 1999; President's Council on Bioethics, 2004)
- the circumstances under which individuals may be enrolled in research (see for example, National Bioethics Advisory Commission, 1998)
- what societies purportedly owe to individuals with regard to health and well-being (see, for example, UNESCO, 2005).

In matters as serious as these, it is understandable that someone might wish to establish "expert consensus" or "expert recommendations" and use the force of expertise to drive social processes. Uncertainty is disconcerting, and adhering to guidelines recommended and agreed upon by experts may ease uncertainty. For further discussion of expertise in bioethics, see Rasmussen (2005). For further discussion of the role secularization has played in the development of bioethics and the notion that bioethicists are experts who can offer guidance, see Engelhardt (2002). Yet, as this essay demonstrates, who counts as a bioethicist and what makes them bioethicists is ill-defined. We should be suspicious of claims such as "bioethicists recommend that. . . ." At best, such statements mean that "some people who are called bioethicists (or who call themselves bioethicists) recommend that. . . ." The latter surely is a weaker statement than the former and suggests far less expert consensus than otherwise might be assumed. And, as this essay demonstrates, it is far from clear what those who bear the title "bioethicist" share in common, which raises questions about what it can mean to be an expert in bioethics.

7.1 A Particular Vision of Bioethics: The One

In the bioethics literature, we often find either (1) an assumption that bioethics is a robust field or discipline whose members share core values, beliefs and goals, or (2) a desire to demarcate the field or discipline and establish a shared understanding of what bioethics was, is, and ought to be in the future. While diversity within bioethics and among its so-called practitioners is often praised, certain types of diversity can be seen as a challenge to some visions of what bioethics is or ought to be. Later in this chapter we argue that the effort to articulate a vision of bioethics as a field or discipline with shared values, beliefs, and goals either presently or in the future is misguided. Here, we document some examples of efforts to articulate a robust core purpose or set of values that define bioethics.[3]

Loretta Kopelman has played an important role in bioethics for several decades, providing insightful analyses not only of specific topics and cases but examining the bioethics enterprise itself. As inaugural president of ASBH the year the organization was born from the combination of the American Association of Bioethics, the Society for Bioethics Consultation, and the Society for Health and Human Values, she posed the following problem:

> What does it mean to say we are "one field," when we come from so many different disciplines? Our members include those trained in law, medicine, nursing, allied health, literature, history, philosophy, religious studies, dentistry, and ministry. How can something be one field and yet be so overwhelmingly interdisciplinary? . . .
>
> "Our field". . . .has no set of common examinations leading to a single degree. We come from different academic homes, each with distinct languages, methods, traditions, core curriculum and competency examinations (Kopelman, 1998, p. 357).

She answered her question of what makes bioethics and humanities "one field" by teasing out the "framework assumptions" of the field. These framework assumptions ultimately shape how practitioners understand the duties or commitments the field begets. She identified six features that unite the field: bioethicists (1) work on set of well-defined, human-condition problems that drive the field; (2) use interdisciplinary approaches on these problems; (3) use cases and practical reasoning to understand and solve moral dilemmas; (4) use Deweyian-inspired pedagogical methods and goals, in order to teach problem solving; (5) aim to find morally justifiable resolutions or solutions to ethical problems; and (6) seek interdisciplinary collaboration, scholarship, service, and teaching (1998, p. 358). Although these framework assumptions are broad, they reflect an attempt to demarcate bioethics and articulate what bioethics is and is not. Kopelman offers an analysis that leads to the conclusion that bioethics has a core set of method and goals. These methods and goals can be interpreted broadly or narrowly. For example, to "solve" moral problems might mean that we strive to find the moral truth and implement morally correct actions or it might mean that we try to find peaceable solutions to conflicts. If these framework assumptions are interpreted broadly, they may offer an accurate description of the present-day bioethics enterprise. Others have proposed more robust and, according to our discussion later in this chapter, less plausible accounts of the core purpose, goals, and values that frame bioethics.

Edmund Pellegrino defends a different understanding of what bioethics once was and again ought to be. Pellegrino laments that bioethics has grown nebulous in its attempt to accommodate and integrate a patchwork of disciplines and professions and it has strayed too far from an emphasis on the normative. As he observes, "Bioethics, today, is a multi-disciplinary, inter-professional, and multicultural enterprise," and "has embraced virtually all of the social and behavioral sciences as well as law, politics, and economics" (Pellegrino, 2000, p. 656). This inter-disciplinary approach has had the unfortunate effect of diluting the "normative thrust" of the ethical component of bioethics. Moreover, Pellegrino is concerned that post-modernism has destroyed the "ethics" in which bioethics is supposed to be located. That is to say, the traditional quest of ethics to seek norms and acknowledge universal values, such as truth, goodness, rightness and so forth, has been eroded in contemporary bioethics. He argues that as a result of bioethics' attempt to accommodate multiculturalism, secularism, political liberalism, and postmodernism, the field "has become a procedure for resolving 'value' conflicts, whether those values are moral values or not" (p. 656). He similarly blames the post-modernist attack on Enlightenment reason as an avenue to moral truth for the contemporary abandonment of moral norms within the field. According to Pellegrino, "With its central pediment gone [reason], the Enlightenment project of a religion-free, metaphysics-free ethics ends in the destruction of normative ethics. The only recourse to is negotiation, since reason itself is now discredited...What sought is now what 'works' or what is 'useful,' i.e., what is justifiable to the parties making the decision, irrespective of whether or not the conflict resolution is true or good " (2000, p. 657). Pellegrino makes strong essentialist claims regarding what bioethics *ought* to do or to be to defend itself from this ethical crisis. First, it will need to recover, from philosophical and theological ethics, the normative principles, axioms, guidelines, and so forth. Second, it will need to continue to engage other disciplines, but take care not to submit wholly to any of them.

> A central challenge for bioethics is how to preserve and recover its ethical credibility, that is to say its normative content, one with universal applicability and acceptability. This challenge cannot be med by abandoning ethics as it has been understood traditionally and replacing it with other disciplines in the humanities or social science. Nor is the seductive substitute of "discourse ethics" likely to recover normative ethical content (Pellegrino, 2000, p. 658).

Finally, "Ethics will need to recapture its identity as a discipline characterized by a method of analysis and a body of literature whose specific end is ascertaining moral right and wrong, good and evil in human conduct...For ethics to remain at the center of bioethics, it need not require a rejection of other disciplines...However, when they replace ethics, or attenuate its normative thrust, they destroy the enterprise to which they genuinely contribute" (Pellegrino, 2000, p. 661). Pellegrino offers a very precise definition of the scope and purpose of bioethics. This definition is characterized mainly by the fact that he requires that other disciplines within the field ultimately submit to the discipline of ethics. According to his argument, what is essential to bioethics is ethics itself. That is, bioethics should be a normative enterprise aimed at universal moral truths.

Other commentators have focused on the social purpose of bioethics and have defended a view of bioethics as an enterprise that should focus more on social problems and concerns and less on issues of individual morality. Underlying these discussions seems to be an assumption that particular ways of organizing society and responding to social problems should inform the bioethics discourse. To understand certain problems as social problems or as matters of social justice itself presumes that persons engaged in the bioethics enterprise do or should hold particular moral or ideological views. Albert Jonsen's discussion of the social responsibilities of bioethics serves as one example of a proposal for the future development of bioethics that presumes bioethics is or should be a unified field framed by particular norms or values (2001). That is, he calls for bioethics to incorporate discussion of social responsibility into its discourse and to highlight, rather than marginalize, the responsibilities of individuals in their social environments. Does Jonsen hold that there is or could be a shared understanding of social responsibility that emerged from our grounded bioethics discourse? Must all discourse engage this notion of social responsibility as a necessary condition for qualifying as bioethics discourse? If bioethics is diverse in some of the ways we consider later in this chapter, any attempt to suggest that persons engaged in the bioethics enterprise must share and emphasize social responsibility is problematic.

Renee Fox argues for an even more robust conception of bioethics as an enterprise that should focus on the social. She argues that bioethicists (and thereby, the field in general) ought to focus more on underlying issues of social justice. She remarks that bioethics:

> has devoted more attention to questions pertaining to the forgoing of life-sustaining treatment of infants in the neonatal intensive care unit than to why a disproportionately high number of extremely premature infants, of very low birth weight, with severe congential abnormalities, cared for in such a unit, are born to mothers from deprived socioeconomic backgrounds... [I]t is more energetically involved in ethical issues associated with physician-assisted suicide that have captured public, media, and legislative attention than with the everyday challenges to physician ethics posed by the changing health care delivery system in the United States, especially by the proliferation of marketplace-oriented managed care organizations and health plans (Jonsen and Fox, 1996, p. 5).

Fox criticizes bioethics for focusing on medical decision-making rather than the underlying social issues that may (or may not) contribute to the necessity of the decision in the first place. This assumes that bioethics ought to encompass a very broad social role. While there is nothing wrong with this assumption per se, the implication that bioethics is wrong not to fulfill her ideal vision of the practice (i.e., to focus a majority of its attention on socioeconomic issues) is problematic. It presumes that persons engaged in the bioethics enterprise are or should be engaged in one mission, share one set of interests and concerns and goals, and one understanding of facts. That is, to see certain problems as matters of social justice already is to presume a particular moral or ideological standpoint. To presume that those engaged in bioethics do or should share that standpoint is problematic.

Other commentators have focused even more explicitly on the obligation of persons engaged in the bioethics enterprise to serve as agents of social change. These views often have been expressed as part of the "taking stands" debate. The "taking stands" debate refers to a conflict over the proper role of bioethics, and specifically of ASBH. According to Armand Matheny Antommaria, the debate arose within ASBH after two events: (1) in 1999, the Board of Directors was asked publicly to protest the treatment of one of its members, whose candidacy for promotion at Medical University of South Carolina was denied following her controversial court testimony against the University's policy on substance abuse during pregnancy; and (2) in 2000, American Academy of Medical Colleges asked the organization to endorse a statement on responsible conduct for research (Antommaria, 2004, p. 24). These events instigated a fierce debate within the organization, and as a result, the Bylaws were amended to allow ASBH to "adopt positions on matters related to academic freedom and professionalism" (p. 24). However, the conflict supersedes ASBH bylaws: the question of "taking stands" is relevant to the entire field (e.g., in directing public policy) as well individual practicioners (e.g., in providing clinical consultations). Throughout this debate, different visions of bioethics and of a bioethics organization have emerged. As Rasmussen explains:

> One side of this division sees bioethics as providing neutral guidance (e.g., to the public in interviews or to patients and families in consultation) in medical ethics, on this account, it is the job of bioethics to shed light, but not direct choice. The second side envisions bioethics not as neutral, but rather as specifically responsible for pursuing a particular moral perspective. This varies from advocating for individuals... to advocating for the moral rightness of specific positions...to operating as an advocate for social justice (Rasmussen, 2005, p. 97).

Some contributors to this debate have maintained that as the major organization that represents bioethics in the United States, ASBH should take public stands on numerous issues and, in this way, attempt to be an agent of social change. Often, these calls for public stands have included not only a call to take a stand but to take a particular stand. For example, Erich Loewy, argues:

> Every "specialty society" I know favors—and most have put forth—a plan for universal access....Ironically, the only society dealing specifically with healthcare matters that has refused to take a stand is, of all things, the American Society for Bioethics and Humanities (ASBH) and its 3 historically separate predecessors. For at least 20 years (until I finally resigned in protest several years back) I used to rise yearly at these societies' general meetings to suggest taking a stand on what I have long called "framing conditions"—"framing" because without them little else makes sense. The excuse has been that our bioethics societies' constitutions expressly forbid us to take such a stand...The members voted that we would speak out, but only on matters such as academic freedom, tenure, etc. In other words, we would speak out on only those matters that directly concerned us personally. The vote was purely self-interested, and one that clearly expressed the sentiment "I've got mine, the devil with you." This is why, after 20 years of pleading with the society to take a stand on poverty, hunger, lack of opportunity to develop one's talents (all closely connected with the incidence of disease), and access to medical care, I resigned. To be a bystander to conditions that could in fact readily be changed is to make oneself guilty as an accessory to the fact—precisely what the deafening silence of academics (by no means only in medicine) in Nazi Germany did (Loewy, 2007, p. 41).

Thus, not only is Loewy appealing to a notion of a commonly understood good inherent to bioethics, he is suggesting that to deny such a good is highly unethical. This commonly understood good, further, ought to be championed and explicitly expressed by ASBH. Loewy presumes that the public face of bioethics (ASBH) *ought* to act as an advocate for particular social and economic policies he presumes reflect the morally right choice. Failure to hold these views and publicly advocate for them is, for Loewy, akin to shirking a moral duty (and doing so is as egregious as contributing to the spread of Nazism through silence).

7 The "s" in Bioethics: Past, Present and Future

The above are but a few examples of arguments that reflect or call for a one-ness in bioethics – a robust core, purpose, function, or set of values and norms that should inform or drive bioethics. In the remainder of this chapter, we argue that bioethics is plural and diverse in a number of important ways and that the goal of identifying a robust core that unites bioethics is elusive. We turn now to examine some of ways in which bioethics is many and not one.

7.2 The Bioethics Enterprise: The Many

Evidence of pluralism in the bioethics enterprise abounds, and questions of who is a bioethicist and what bioethics is have plagued presidents of ASBH since the Society's inception. As Tod Chambers noted in his address as the incoming president of ASBH in 2007, many incoming presidents have explored the divisions in bioethics and diversity among ASBH members (see, for example, Kopelman, 1998; Montgomery, 2002; Derse, 2005; Wolpe, 2007). It is important to document this pluralism and identify some of the many "we's" who constitute the bioethics enterprise because these differences are relevant to ongoing discussions about the education of bioethicists and the life of an organization such as ASBH. There are many ways one could draw lines that divide the bioethics enterprise to demonstrate the significance of the "s" in "bioethics." We consider four axes of difference that affect the education of future bioethicists and the life of a bioethics organization such as ASBH.[4] These are: (1) disciplinary differences, (2) functional differences, (3) subfields or areas of specialization within bioethics, and (4) religious, ideological/moral and cultural differences.

7.2.1 Disciplinary Differences

It is often stated that bioethics is interdisciplinary. This interdisciplinarity can mean at least two different things. First, bioethics is informed by the work of multiple disciplines and to do their work, bioethicists must be in dialogue with the literature of multiple disciplines. As Kopelman observed, "justifiable solutions to ....difficult problems about the human condition transcend our individual disciplines and require interdisciplinary approaches" (Kopelman, 1998, p. 359). For example, to understand and contribute to discussions about access to health care and health insurance, one must recognize the significance of contributions from political philosophy, religious studies, public policy, law, and psychology, among others.[5] Second, the observation that bioethics is interdisciplinary can refer to the fact that bioethicists come from different disciplines and many have professional identities other than being bioethicsts. There are bioethicists trained in philosophy, theology, religious studies, law, psychology, medicine, nursing or other health professions. Kopelman has argued that these factors make bioethics a second order discipline, one that "is a 'higher' type of entity constructed from, but not reducible to, a 'lower' type." She observes that, "Successful bioethicists come from many professions and disciplines and our different roots nourish both the individual bioethicist and the field of bioethics..." (2006, p. 602). The observation that "successful bioethicists" have different professional and disciplinary homes presupposes knowledge of who is a bioethicist, an issue about which there significant disagreement among persons working in and observing the bioethics enterprise. (See Kopelman, 2006, for an insightful analysis of the issue of who "counts" as a bioethicist.) For example, although a person trained in social work might consider himself a bioethicist and do some of the work typically associated with being a bioethicist, such as clinical ethics consultation, some might hold that he lacks the relevant training in ethics – either philosophical or religious – to be considered a bioethicist. The role and significance of disciplinary homes varies from person to person engaged in the enterprise of bioethics. In some cases, individuals' alternative professional identities may have lost significance over the years, e.g., bioethicists with a Ph.D. in theology might work in a bioethics department at a secular university and rarely if ever engage in any religious or theological discourse. For others, these alternative professional identities remain important, even dominant. In some cases, individuals whom others consider to be bioethicists do not identify themselves as bioethicists. During a visit to Saint Louis University in the Fall of 2006, James Childress indicated that he does not consider himself a bioethicist. Many western bioethicists recognize him as a major figure in the birth and ongoing development of contemporary bioethics.

Regardless of how far one casts the net in labeling individuals bioethicists rather than merely recognizing their work as pertinent to bioethics, one is likely to include persons who were not specifically trained as bioethicists, including some who do not have formal training in ethics. We also may disagree about whether there are particular contributions from different disciplines bioethicsts must understand to be considered bioethicists. Some hold that if we had a credentialing/certification process for bioethicists, then confusion about who is a bioethicist and what one must know to be one would dissipate. After all, only those who were credentialed would be bioethicists. One difficulty is that we disagree about what constitutes expertise in bioethics (See Rasmussen, 2005), and an understanding of expertise would be necessary to develop a meaningful credentialing process.

Given (a) the wide range of issues addressed in bioethics, (b) the range of disciplines engaged in the bioethics enterprise, and (c) the disagreement over what it means to be a bioethicist and whether there are any core areas of knowledge all bioethicists must demonstrate, skills they must possess, and so forth, we should expect to continue to see disciplinary diversity for quite some time (if not indefinitely). There is a sense among some that anyone in any discipline that is related to or the object of discussion in bioethics and who addresses issues related to bioethics may claim to be a bioethicist.[6] This has significant implications for how those of us involved in educating future bioethicists plan curricula and how anyone in the field judges the background and training of new or would-be bioethicists. What should we require of our students? Should they have an advanced degree in another relevant field? If so, which fields? What courses should they take and from whom? What knowledge and skills should they demonstrate? One of us has argued elsewhere that "what it is to be a bioethicist and possess bioethics expertise depends in part on the context in which such expertise is sought" (Iltis, 2006, p. 633; see also Iltis, 2005). For example, a Roman Catholic health care institution will understand bioethics expertise as including knowledge of natural law ethics as interpreted in the Roman Catholic tradition as well as Church teachings on bioethics matters. A secular or Jewish institution will not require or even necessarily want a person with such expertise in a bioethics position. At the same time, "[b]ioethicists share in common an ability to gather, synthesize, and integrate relevant data that enable them to address a range of concerns related to health care practice and delivery, health policy, biotechnological developments, and biomedical and behavioral research" (Iltis, 2006, p. 637). The education of bioethicists should be informed by the expectation that bioethicists hold this ability to seek out, gather, understand and synthesize data from different disciplines to analyze specific issues. A number of disciplines make important contributions to bioethics, including law, medicine, nursing and other health sciences, philosophy, biology, chemistry, physics and other hard sciences, the social sciences including sociology, anthropology and economics; psychology, health care administration, public health and epidemiology, literature, religion, philosophy and other humanities. As a result, a bioethics education should foster an ability to (1) determine which contributions from which disciplines will be important for examining an issue and (2) read, understand, and interpret those contributions, and (3) synthesize various contributions to analyze issues and articulate one's analysis and conclusions. (See Iltis, 2006 for further discussion.) This does not resolve many important questions, such as where the emphasis should be or how these abilities should be fostered or how skills and knowledge should be demonstrated.

Interdisciplinarity also has implications for the life of an organization such as ASBH. First, if membership in an organization such as ASBH is contingent upon being a bioethicist (or medical/healthcare humanist)[7] or being a member of such an organization renders one a bioethicist, then membership criteria must be established and such criteria must reflect what it is to be a bioethicist. If membership in ASBH is open and does not establish professional status, as currently is the case, this issue is not relevant. However, there is a risk that individuals may think membership in such an organization confers more than it does. An organization for an interdisciplinary or multidisciplinary enterprise may face difficult questions about how to hold organization-wide conferences. There will be disagreements over what counts as a useful or good presentation, what kinds of themes should receive program time and so on. If bioethicists' professional identity is further specified through a code of ethics or certification/credentialing process, interdisciplinarity will complicate decisions about what should be included in a code as well as what skills, knowledge, and expertise individuals must demonstrate to be bioethicists.

7.2.2 Functional Diversity

Depending on which account or accounts of the history of bioethics one accepts, the early work of people who eventually would be called bioethicists included making or advising on clinical decisions such as access to dialysis (Jonsen, 1998); evaluating new scientific and medical developments such as organ transplantation and brain death (Veatch, 2005); responding to unethical human research (Rothman, 1991)[8]; and regulating science through the responsible science movement (Stevens, 2000).[9] Some scholars identify medical ethics as an important precursor to contemporary bioethics, with some emphasizing the tradition dating to Hippocratic medicine (Jonsen, 1998, 2000) and others focusing on the 18th and 19th century work of John Gregory and Thomas Percival (McCullough, 1998). The books, articles, and other materials in the Roman Catholic manualist tradition aimed at instructing physicians and nurses on medico-moral matters (see, for example, Kelly, 1949–1954 and 1958; LaRochelle and Fink, 1947) also are part of the history of bioethics. At one time, clergy and moral theologians offered authoritative guidance to help individuals understand ethical problems in medicine and make moral decisions. This focus on ethical guidance for practical decisions is reflected in Callahan's recommendation during the early days of contemporary bioethics that "the discipline of bioethics should be so designed, and its practioners so trained, that it will directly – at whatever cost to disciplinary elegance – serve those physicians and biologists whose position demands that they make the practical decisions" (Callahan, 1973, p. 73).

Today, some bioethicists engage in this activity of providing normative advice on medical moral matters, a connection not lost on Engelhardt, who has explored the history of contemporary bioethics in "The ordination of bioethicists as secular moral experts" (Engelhardt, 2002). As bioethics has evolved, regardless of where one thinks bioethics originated, individuals involved in bioethics have engaged in an expanded array of activities. These include clinical ethics consultation and/or mediation, membership on institutional ethics committees addressing matters of clinical and/or organizational ethics, education of health care professionals, membership on institutional review boards or research ethics committees (IRBs or RECs),[10] scholarship (both theoretical and empirical research), teaching students in the health professions, research ethics consultation (de Melo-Martín et al., 2007; Cho, 2008), communication with the press, involvement on government commissions and panels such as Institute of Medicine panels and presidential commissions on bioethics, consulting with attorneys or offering expert testimony in the courts, and providing ethics consultation to corporations such as pharmaceutical companies. Often, no single bioethics function commands all of a person's time but rather a combination of some of these and possibly other activities, such as providing patient care in the case of clinicians or teaching undergraduate philosophy students in the case of philosophers, constitutes the work of a bioethicist.

Not all bioethicists engage in all of the above-named activities, and some of those functions – especially offering expert testimony in courts and providing corporate consultation (Brody et al. 2002a, b; DeVries, 2004; Youngner and Arnold, 2002;

DeVries and Bosk, 2004; Elliott, 2001; Rasmussen, 2005) – have been the subject of controversy. Perhaps no potential function of bioethicists has been more controversial than the activist role. Some hold that activism is a function of bioethicists (Parker, 2007; Marshall, 2007). Some of the qualities of a good activist, e.g., being able to engage people's emotions so that they accept definitive conclusions about what we ought to do even when there are conflicting data so as to bring about social or political change, are incompatible with what others take to be the role of bioethicists, such as scholarship or serving as public intellectuals. Parsi and Geraghty have argued that bioethicists should function as public intellectuals. The bioethicist as public intellectual "synthesizes information from various sources, integrates it into a coherent message, and addresses a broad audience of readers, listeners and viewers. . . .They help individuals think more deeply about the far-reaching effects of medicine and technology in our lives" (2004, p. W22). This goal can conflict with the goals of activists who aim to motivate particular beliefs and actions and to garner support for particular social changes.

A further point of controversy in the activism debate emerges when we recognize the diversity of moral/ideological, cultural, and religious views bioethicists hold, an issue discussed further below. Some who have urged that bioethics is or should be a form of activism have suggested that there are obvious matters for which bioethicists should be activists, such as social justice (where a particular understanding of social justice is assumed) (Parker, 2007). Even if one accepts activism as a function of bioethicists, there will be disagreement over the scope and focus of such activism. For example, while some might assume that it would be appropriate for a bioethicist to be an activist who defends a woman's right to choose abortion but not to be a pro-life activist, others will hold the opposite view. Just as some think the most obvious form of vulnerability from which bioethicists should attempt to rescue people is lack of health insurance, others will see that the most obvious and extreme form of vulnerability is being in a position in which others legally have the right to kill you and you have no right or ability to defend yourself. There is no way through reason alone to explain why the former is seen as so obviously the obligation of bioethicists and the latter is not. Similarly, while some will argue that bioethicists should fight for universal, tax-funded health coverage, others will hold different positions. The issue of ideological/moral, cultural, and religious diversity among bioethicists is discussed below. We note here only that such diversity exists and if bioethicists are to function as activists, there will be multiple visions of the ends they should defend and changes they should seek to bring about.

Insofar as bioethicists will serve competently in various capacities, they require different training and skills, and they will be evaluated using criteria appropriate for each function. As we consider educating bioethicists, we must ask: are some functions core to bioethics in that anyone who will be called a bioethicist must be trained and demonstrate skills in those areas while others are peripheral or optional? If so, what are they, how should individuals be educated, and what constitutes demonstrations of appropriate knowledge and skills? If not, how do we decide what training to offer and what skills to test? In arguing that we should not develop a code of ethics for bioethicists qua bioethicists, John Lantos reflects on the functional diversity of bioethicists and concludes that there is no core activity of bioethics (2005). If there is no core, does anyone sufficiently trained to perform any one function count as a bioethicist, or must one possess the skills and knowledge to perform a critical number of functions? One's answer to this question will depend on what one holds are the essential and legitimate roles for bioethicists. Additional questions include: what does it mean to be sufficiently trained for any of these roles? How should one demonstrate skills or knowledge?

An organization that includes bioethicists with different – sometimes radically different– functions will be affected by this diversity in multiple ways. First, insofar as such an organization wanted to develop a code of ethics for its members, it would have to recognize that obligations of bioethicists acting in certain capacities would not necessarily apply to bioethicists acting in other capacities. In the case of ASBH, the situation is even more complicated, in that membership in the organization neither grants nor requires that one be a recognizable member of any profession or have any particular credentials. Robert Baker has been a leading voice in the movement to develop a code of ethics for bioethicists and has argued that a code should be broad and relevant to all bioethicists (2005). Others have expressed concerns about formulating a code for bioethicists that is not specific to bioethicists working in particular capacities. Kipnis articulates this concern, noting that scholars who teach in a university setting have obligations and likely face ethical challenges different from clinical ethicists who see patients (Kipnis, 1997, p. 49). Lantos observes that some bioethicists, namely those engaged in clinical ethics consultation, "perform some functions that are both superficially and structurally similar to those of these other healthcare professionals" who have codes of ethics against which their behavior may be judged (Lantos, 2005, p. 46). Bioethicists "also perform many functions that are more similar to those of our colleagues in traditional academic disciplines" such as writing papers and teaching, for whom a separate code of ethics seems unnecessary as they already have codes of ethics (e.g., through the American Association of University Professors) (Lantos, 2005, p. 46). Lantos argues that there is no core activity of bioethicists, at least not in the same way there is a core activity of medicine or nursing (taking care of patients). Even though physicians, for example, may be involved primarily or exclusively in non-patient care activities, they were trained to care for patients and the codes of ethics that govern physicians primarily focus on issues related to this core activity. Lantos concludes that "[a] code for a field without a core is likely to be either divisively substantive – focusing on one set of activities and the norms requisite to that activity to the exclusion of other important activities – or inclusively bland, describing norms of behavior that are universal and vague" (Lantos, 2005, p. 46). Miller (2005) agrees that any code of ethics should be narrow and apply to individuals engaged in specific activities: bioethicists acting in any capacity that already is subject to a code of ethics do not need an additional code, e.g., professors are subject to the AAUP code, physicians to the AMA code and so on. He notes that in his entire career, he "can't think of a single instance" in which a professional code for bioethicists would have made a difference for him (Miller, 2005, p. 51).

The range of functions bioethicists perform also has implications for how a society, such as ASBH, conducts business and how journals evaluate and select contributions. Taking into account the breadth of what "counts" as bioethics requires many often controversial decisions. For example, meeting time discussing clinical ethics consultation is well-spent for some but not others. Some will hold that papers should be scholarly and because of disciplinary differences there will be different standards for determining what is scholarly, while others will want an emphasis on what they perceive as more practical presentations and essays (e.g., how to manage difficult clinical ethics consultations). Some may want to spend time hearing about the most effective techniques for speaking persuasively and successfully bring about social change, while others will think the most appropriate way to spend time at a meeting is exploring nuanced arguments about a new medical technology. This does not mean that it is impossible to hold meetings and publish work that meet the needs and interests of different individuals, but decisions must be made about which areas will get how much attention, which will receive none, and how contributions will be evaluated.

7.2.3 Sub-fields/Sub-specialization

To date, there remains disagreement over what the word bioethics means and the scope of concerns that fall under this term. The origins of the term "bioethics" has been traced to Van Rensselaer Potter, André Hellegers, and Sargeant Shriver, each of whom envisioned something different (see Reich 1993, 1995, 1999). Kopelman (1998) argued that one feature of bioethics is that persons engaged in bioethics:

> work on trying to solve a cluster of problems about the human condition, many of which are urgent, momentous and interrelated. We address topics related to human vulnerability such as sickness, suffering, birth, death, and the pursuit of health, and the avoidance of pain, loss, and injury ... We consider issues such as abortion and personhood, as well as how we ought to treat people facing death who want to have control over their lives (Kopelman, 1998, p. 358).

The sub-specialization or sub-fields of inquiry within bioethics, such as clinical ethics, research ethics, organizational ethics, neuroethics, nursing ethics, and medical ethics reflects this breadth. Some see bioethics as stretching further, to include areas of inquiry such as environmental ethics and public health ethics.

The range of concerns addressed within the bioethics enterprise has important implications for educating bioethicists. What should be part of a bioethics education? When should sub-specialization begin? Should all study the same core areas? What should the core include? For example, should all bioethicists demonstrate a degree of knowledge about philosophical and religious ethics, medicine, law, and public policy?

Sub-specialization also leads to in important issues for an organization such as ASBH. As anyone involved in ASBH mostly likely knows, there have been debates over what areas "deserve" an affinity group [i.e., a sub-group of ASBH members meant "to foster scholarship, education, and collegiality on a topic of specific common interest to its members" (ASBH Bylaws Article X, section 1), such as organizational ethics, clinical ethics, law and bioethics, spirituality and bioethics]. Similarly, there may be debates about how much program time to allocate to various subfields, what topics are appropriate for plenary or keynote talks, pre-conference workshops and so on. These debates reflect the many areas compassed by the bioethics enterprise.

7.2.4 Religious, Cultural and Moral/Ideological Pluralism

Individuals who consider themselves to be bioethicists or whom others might recognize as being part of the bioethics enterprise hold a wide range of religious, moral/ideological, and cultural commitments. For example, there are atheists, agnostics, Christians, Jews, Hindus, Muslims, Bhudists, Daoists and others. Among those who hold religious commitments, in some cases religion plays a role – perhaps a major role – in their work, while for others such religious commitments are set aside and do not shape their bioethics work explicitly. Within any group identified above, we should expect to find people who hold different positions on some bioethical issues despite the fact that they share some common identity. For example, among Christian bioethicsts, there are Roman Catholics, Lutherans, Episcopalians, non-denominational Christians, Baptists, Orthodox Christians, and so forth. Even among these sub-groups we should expect differences, such as more liberal and more conservative Roman Catholics. Among Jewish bioethicists, there are Orthodox, Conservative, and Reformed Jews. These differences are real and significant. They have implications for understanding many issues in bioethics, including the nature and respect owed to unborn human life as well as to the elderly or disabled, the proper relationship between humans and animals (e.g., the use of animals in research), appropriate end of life care, and the use of medications or particular medical interventions, to name only a few. Some people will hold moral/ideological commitments that they do not identify as religious commitments but that nevertheless shape their understanding of issues in bioethics. For example, one can imagine a person who is committed to animal rights and opposes particular types of animal research or one who is committed to equal access to health care for ideological reasons that are not based on religious commitments.

In addition to religious and moral/ideological differences, there are people whose cultural backgrounds differ in ways relevant to bioethics. For example, some cultures ascribe different levels of significance to the family, e.g., for some of us it is unthinkable that major health care decisions would be made by ill, elderly patients alone rather than in consultation with family members. For others, the authority of individuals over themselves is of primary importance. Cultural differences may affect the extent to which patients or families will question physicians' decisions or recommendations, seek additional opinions or make health care decisions. Cultural differences may affect the extent to which individuals trust the state and have confidence in state-controlled health care.

7 The "s" in Bioethics: Past, Present and Future

Bioethics is a global endeavor. There are people all over the world engaged in activities they refer to as bioethics or that others might consider bioethics. A substantive literature has emerged in, for example, China (see Li and Wen, 2010; Chen and Fan, 2010; Wang and Wang, 2010; Chen, 2007; Fan, 2004, 2007), Japan (see Hoshino, 1997); the Philippines (see Alora and Lumitao, 2001); Latin America (see Pessini et al., 2010); India (see Kumar, 2006; Chattopadhyay and Simon, 2008); and Africa (see Tangwa, 2000; Gbadegesin, 1993). The differences that emerge not only within cultures and regions but between them add further to the plurality of bioethics (see Tao, 2002; Engelhardt and Rasmussen, 2002).

If, as Engelhardt (1996) has argued and as experience suggests, there is no way through reason alone to resolve these disagreements or to identify a substantive shared morality because competing and incompatible positions ultimately turn on initial premises that cannot be established or falsified through reason alone, then there is no neutral bioethics (see, also, Engelhardt, 2006). The significance of this should not be underestimated. Non-belief in a particular religious principle, non-acceptance of a particular cultural norm, or a decision to set aside religious convictions in analyzing issues does not reflect neutrality. In other words, to accept embryonic stem cell research because one rejects all religious commitments that recognize embryonic life as human life that is not to be destroyed is as much a substantive position as holding the embryonic stem cell research is illicit because it involves the intentional destruction of human life. To reject a position is no doubt to adopt another position, and simply because one position rests on religious assumptions and another rests on secular assumptions renders neither one more believable or acceptable to those who do not share the initial premises. As a result, the religious, moral/ideological, and cultural differences we observe among persons engaged in the bioethics enterprise merit our consideration. We should not think we see unity and community where they are absent.[11]

The religious, moral/ideological, and cultural pluralism among persons who are part of the bioethics enterprise suggests that there may be disagreement about the appropriate content and scope of bioethics education, the methods of inquiry that should be employed to explore issues in bioethics, and the standards that define knowledge and truth. For example, what ethical theories should be studied? How should the role of religion in bioethics be framed? What role should culture play in bioethics? What does it mean to respect culture? These differences will affect the life of any organization that attempts to encompass bioethicists writ large. If said organization does not wish to identify itself as representing a particular niche of bioethicists – those who hold specific religious or moral commitments, for example – then the organization must be prepared to acknowledge moral/ideological, religious and cultural diversity as part of its very identity. In developing meeting programs, for example, decisions will have to be made about which perspectives merit how much program time and which merit none, which should be featured as keynote presentations and so on. Such an organization cannot hope to take substantive moral positions on matters in bioethics if those positions are to be portrayed as representing the organization's members' views or definitive normative positions on bioethics matters.

The possibility of having a bioethics organization take substantive positions on topics in bioethics has been under discussion in ASBH for a number of years (see our discussion of the "taking stands" debate earlier; see Antommaria, 2004). Unless and until ASBH defines itself as a particular type of organization that has a specific moral/ideological, religious, or cultural starting point, it is difficult to imagine where the normative authority for its positions would lie and where the content of its substantive positions would originate. Today, ASBH is "an educational organization whose purpose is to promote the exchange of ideas and foster multi-disciplinary, inter-disciplinary, and inter-professional scholarship, research, teaching, policy development, professional development, and collegiality among people engaged in all of the endeavors related to clinical and academic bioethics and the health-related humanities" (ASBH Bylaws, Article 3, section 1). Insofar as this is the purpose of the Society and its members claim to value diversity (see below) rather than to adopt a particular religious, moral/ideological, or cultural standpoint, it is difficult to understand why one would think that the Society should assume the authority to make claims about various topics in bioethics.[12] The idea that a professional organization could draft moral position statements when it professes no particular normative lens and when its very purpose is in part to explore those lenses is odd. No doubt, some will be tempted to note that many other professional organizations that do not explicitly profess to hold a particular cultural, religious, or moral/ideological standpoint do make substantive moral claims and that ASBH should do so as well. Here one would do well to ask whether those organizations that adopt substantive positions despite having no particular moral lens are right to do so. One might hold that bioethicists also should form organizations the purpose of which is to make such pronouncements, but to suggest that the current goals of ASBH are either not worthy or insufficient for a meaningful organization seems inappropriate.

Even though ASBH was not founded as a political organization and it does not identify itself as espousing or representing a particular religious, moral/ideological or cultural vision, some members appear to assume a shared moral/ideological vision exists. In the fall of 2006, Paul Root Wolpe addressed the ASBH annual meeting as incoming president and, as is customary, published a summary of his comments in a subsequent ASBH newsletter. He observed:

> We believe that we at ASBH are open-minded, culturally sensitive, ideologically diverse, nondiscriminatory, and committed to equality and social justice, yet, although we are doing better than just a few years ago – our membership fails to reflect this stated commitment to racial, ethnic, disabled, sexual, and other minority representation (Wolpe, 2007, p. 2).[13]

It is interesting to observe that Wolpe spoke of a commitment (1) to ideological diversity and (2) to equality and social justice as compatible and perhaps core commitments of ASBH members. A commitment to equality and social justice is an ideological commitment, and thus if ASBH members are universally committed to this ideology, then the organization is not ideologically diverse (at least on this matter). This is one example of an instance in which ideological diversity among those engaged in the bioethics enterprise is treated as important only to be dismissed.

7 The "s" in Bioethics: Past, Present and Future   139

Imagine if instead of being "committed to equality and social justice" Wolpe had suggested a commitment to respect for human life from conception to natural death. No doubt many people would have been startled to think that an organization committed to ideological diversity also could be committed to respect for human life in this way. The commitment to respect for human life from conception to natural death is no less an ideological commitment than a commitment to equality and social justice. If ASBH does maintain that it is an ideologically diverse organization, this will affect its ability to assert positions on substantive moral issues as if they represented the organization. One wonders whether some of those who maintain that ASBH should take substantive positions on matters related to bioethics hold that view in part because they are confident that they are in the majority and will be able to ensure that their views are reflected in the positions articulated (or not articulated) by the Society. For example, do they hold that ASBH should take substantive positions because they are confident, that ASBH won't take a call to speak out for the vulnerable victims of injustice as a call to make a statement that abortion is murder?

7.3 The "s" in Bioethics Matters

"Mary is a teacher." The statement does not tell us much about Mary. Each person hearing that statement might make certain assumptions about Mary, but our initial perceptions (which most likely are based on our own experiences and background) could be wrong. Mary could be an elementary, middle, high school or pre-school teacher. Mary could teach any number of subjects at any of those levels, e.g., she could be a third grade classroom teacher or an elementary art teacher, and eighth grade science teacher, or a high school physical education teacher and so on. Mary also might be a private piano teacher, or she might be a homeschooler. Perhaps she teaches at the college level, or she teaches photography in a local community center. She could be a Montessori teacher, or she might be trained in constructivist education, or some other teaching philosophy or approach. Maybe she teaches through a local driving school, or teaches in her church's education program. The statement "Mary is a teacher" tells us nothing about her education, her area(s) of expertise (if any), her politics, her religion or lack thereof and so on. The pluralism and diversity we find in bioethics suggests that "Mary is a bioethicist" similarly tells us little about Mary. She might hold a Ph.D. in philosophy and teach at the undergraduate or graduate level; hold a Ph.D. in religious studies and teach medical students and residents; hold a J.D. and teach bioethics in a law school; be a nurse in a hospital who does clinical ethics consultations; be a physician who chairs an ethics committee and so on. She might be pro-life or pro-choice, favor embryonic stem cell research or oppose it, support a right to physician assistant suicide or not, believe that health care should be publicly funded through tax dollars or not, favor uncontrolled donation after cardiac death or not, believe we should adopt presumed consent statutes for organ donation or not, believe that parents should be allowed to enroll their children in research posing more than minimal risk with no prospect of direct benefit or not and so on. This should not surprise us and we should expect that, for the foreseeable future, the claim that "Mary is a bioethicist" will tell us very little about Mary.

The reality of disciplinary and functional diversity as well as the tendency toward sub-specialization should come as no surprise given the history or histories of bioethics. The reality of religious, cultural and moral/ideological pluralism among bioethicists also is to be expected in our society. As of 2001, 76.5% of adults in the United States identified themselves as Christian, 1.3% identified themselves as Jewish, 0.5% as Muslim, 0.5% as Buddhist, 0.3% as Hindu and smaller percentages named other religious affiliations, such as Native American, Scientologist, Baha'I, Taoist, Eckankar, and Sikh, among others (Kosminet al., 2001). 14.1% of adults in the U.S. identified themselves as having no religion (including people who identified themselves as atheist, agnostic and humanist). Numerous Gallup Polls (see www. Gallup.com) reflect disagreements among Americans regarding issues in bioethics. U.S. Census Bureau data indicate that the racial and ethnic diversity of the U.S. population is increasing, with certain minority groups seeing significant population increases. We argue that all four types of diversity in bioethics have significance and we should avoid attempts to ignore or eliminate differences.

Disciplinary diversity is evident in the history of bioethics, and few graduate programs exist today at the doctoral level specifically in bioethics Many people involved in the bioethics enterprise who hold doctoral level degrees hold them in theology/religious studies, philosophy, law, medicine and other fields.[14] If the scope of the bioethics enterprise remains as broad as it is today, then input from many different disciplines will continue to be important. An important question that remains to be answered is whether we can and should differentiate contributors to bioethics from bioethicists. There may be experts who contribute to the bioethics enterprise but are not considered bioethicists. Perhaps the term "bioethicist" will apply to no one. Perhaps it will be a very broad term that applies to many different people but itself implies little about what they know, what they have studied, or what they do. Perhaps "bioethicist" will mean different things in different contexts, so employers will need to know what they are looking for and what needs they are attempting to meet when they hire a bioethicist. To say "Mary is a bioethicist" tells us little about Mary. Perhaps we will identify a core for bioethics training that will enable us to establish who is a bioethicist (one trained in the core? One trained in the core and who has passed certain exams?) and who contributes to bioethics but is not a bioethicist. The scope of problems addressed and types of solutions and analyses sought in the bioethics enterprise require multiple areas of expertise. No one has the knowledge and skills to engage all facets of all bioethics topics on his own. To address multiple facts of different topics well requires both experts who do in depth work in particular facets of an issue and individuals who have the skills to bring together contributions from those different areas to analyze problems. It would be the very rare individual who has studied in depth all the disciplines relevant to analysis of individual problems in bioethics and thus provide the knowledge necessary and then synthesizes all of that information. A person who does not have in depth training in all relevant disciplines but has received sufficient training in different disciplines so as to understand their contributions and be able to integrate those contributions to analyze a problem may make a unique contribution to the bioethics enterprise. We should expect that people with in depth training in one area pertinent to bioethics might also develop some understanding of other areas and thus synthesize and integrate the information to address bioethics problems. Perhaps the people who do this work are "bioethicists." Precisely who, if anyone, will be called a "bioethicist" remains to be seen. If the bioethics enterprise is to continue to address a wide range of issues, it will be important to maintain disciplinary diversity. Whether members of all contributing disciplines are considered bioethicists or not, there will remain a need to have contributions from a range of disciplines.

The functional diversity we observe in the bioethics enterprise reflects a basic fact about human beings – we are not all good at everything. Given the range of functions the bioethics enterprise involves, it is pragmatic to encourage some people to provide some services and not others. A good teacher may not be a good clinical ethics consultation or a good policy analyst and so on. A separate question is whether there are some functions that should not be associated with the bioethics enterprise. We set this question aside here and claim only that preserving functional diversity among individuals in the bioethics enterprise is an important way to promote excellence or at least to avoid mediocrity.

Sub-specialization is trend in many areas and sub-specialization in the bioethics enterprise may be a result of the complexity of some of the issues addressed, the need for input from many different disciplines, and the broad range of issues that fall within the scope of bioethics. Limiting sub-specialization could weaken the level of discourse in bioethics.

Finally, cultural, moral/ideological, and religious diversity cannot be eliminated through rational discourse. In the absence of reliable ways to resolve disagreements through rational discourse, we should not presume that it is permissible to dismiss these differences and assume that one cultural, moral/ideological, or religious position may be held out as defining bioethics or representing the shared vision of anyone who is legitimately involved in the bioethics enterprise. Nor should we expect individuals associated with the bioethics enterprise to share cultural, moral/ideological, or religious commitments.

Some contributors to the bioethics enterprise hold that there is a common or shared morality that grounds or can frame bioethical discourse. For example, new natural law theorists Finnis, Grisez, and Boyle hold that the precepts of the natural law are in principle accessible to all through reason. The precepts of the natural law:

> are natural in the sense that they are not dependent for their validity on human decision, authority, or convention. Because of the independence of these factors, natural precepts and principles must be generally accessible to human reason; the critical reflection that is not dependent upon but potentially critical of any particular social enactment or practice is the work of common human reason (Boyle, 2004, p. 2).

The principles and norms of the natural law are "addressed to anyone at all facing a choice upon which the principles and norms bear" (Boyle, 2004, p. 2). In other words, they are universally knowable and applicable. According to new natural law theorists, there are basic, intrinsically valuable human goods (such as truth, peace, life, health, knowledge, play, friendship, religion and aesthetic experience), the "pursuit of which seems of itself to promote persons and bring them together" (Finnis et al., 1987, p. 277). These goods are basic to human flourishing: they are "intrinsic aspects, – that is, real parts – of the integral fulfillment of persons" (p. 277). As such, they "provid[e] the reasons to consider some possibilities as choiceworthy opportunities" (p. 277). Without prior reasons, the basic goods give us "reasons for choosing and acting" (p. 278): one ought to make choices that promote human flourishing and one ought never intentionally to act against any of the basic goods.

It is questionable whether the precepts of the natural law are in principle universally knowable through reason. The basic goods account presupposes an account of human flourishing, and we have no evidence that such an account is universally available through reason. Moreover, Finnis, Boyle, and Grisez develop an account of what it is to act against the basic goods that turns on assumptions not available to all through reason alone. For example, they hold that contraception violates the basic good of life (see Boyle, 1980; Grisez et al., 1988). Yet many would argue that insofar as contraception does not involve the direct and intentional destruction of life, some would argue that to use contraception is not to act against the good of life. Thus, even one who accepts the basic goods account might derive different moral content when relying on reason. Finnis, Boyle, and Grisez recognize this possibility of disagreement and hold that ultimately such disagreements are not deep disagreements. Persons share common moral ground despite divergence on particular matters. Moreover, disagreements among rational individuals can result from moral error: "individuals and communities can fail to know even fairly general and fundamental moral truths. Aquinas thought that people are often ignorant of, or mistaken about, the norms which should guide their choices" (Boyle, 2006, p. 316).

A lengthier assessment of new natural law theories would take us too far astray from the purposes of this essay. However, these brief comments demonstrate that the claims of a morality universally knowable through reason rest on prior, undisclosed assumptions that may not be universally knowable through reason. Not only does the list of basic goods depend on a debatable understanding of human flourishing, but one's interpretation of what it means to respect the basic goods also rests on a series of assumptions that are not knowable and demonstrable through reason alone.

New natural law theorists are not alone in proposing that there is a universally knowable morality accessible through reason. Tom Beauchamp and James Childress, for example, hold that there is a common morality shared by all persons that "contains moral norms that are basic for biomedical ethics" (Beauchamp and Childress, 2008, p. 12).These norms consist of (at least) the four mid-level principles for which Beauchamp and Childress are so well known: beneficence, nonmaleficence, respect for autonomy, and justice. These principles in some sense transcend moral/ideological, religious, and cultural differences and allow individuals to examine bioethics issues and cases through a shared framework. Individual cases require that we interpret and apply the principles in a process known as specification. Beauchamp and Childress are well aware of disagreements about bioethical

7 The "s" in Bioethics: Past, Present and Future

matters among rational persons. They explain that the common morality accommodates such diversity and that persistent disagreement does not reflect poorly on the common morality theory:

> Conscientious and reasonable moral agents understandably disagree over moral priorities in circumstances of a contingent conflict of norms. … Such disagreement does not indicate moral ignorance or moral defect. We simply lack a single, entirely reliable way to resolve many disagreements, despite methods of specifying and balancing. (Beauchamp and Childress, 2008, p. 24).
>
> We can assess one position as morally preferable to another only if we can show that the position rests on a more coherent specification or interpretation of the common morality (Beauchamp and Childress, 2008, p. 25).

That standard for demonstrating the superiority of a moral position reflects a significant problem: one must have an account of what would be a "more coherent specification or interpretation of the common morality" and such an account is not available from the common morality itself. The very basis for specifying or interpreting principles must come from outside the common morality. It is debatable whether a common morality exists at all and that it exists as articulated by Beauchamp and Childress. Even if it does exist as they describe, the framework of shared mid-level principles is of limited value for the purpose of reasoning about how to resolve particular cases or issues. As Beauchamp and Childress themselves note, rational persons can disagree despite sustained critical reflection and dialogue. As Engelhardt has argued:

> …the appeal to middle-level principles will not resolve controversies, but instead highlight their depth. For example, in determining whether and to what extent high-cost, low-yield medical intervention should be provided by the government to the poor, how will an appeal to the principle of justice provide a resolution when the disputants include individuals who are both Rawlsians and Nozickians (i.e., individuals who hold that justice involves providing all with the material conditions for equality of opportunity versus those who hold that justice involves not interfering with the property rights of others)? The differences depend not just on different reconstructions of the same moral understandings, but grow out of foundationally different moral visions. Nor will an appeal to the principle of autonomy resolve disputes concerning the appropriateness of a twenty-eight-year-old quadriplegic committing suicide. Those who understand the principle of autonomy as expressing an overriding right of people to be in authority over themselves will recognize the right of competent individuals to exit this life whenever they want, with whosoever help they may be able to obtain. Those who understand the principle of autonomy as underscoring values associated with liberty may argue that an early death is a loss of many years of enjoying liberty and therefore find such choices not only wrong, but meriting coercive restraint (Engelhardt, 1996, pp. 57–58).

Despite the differences between the new natural law proposed by Finnis, Boyle, and Grisez and the common morality approach proposed by Beauchamp and Childress, they suffer from a similar problem. The moral frameworks can be universal in only a very thin and morally uninteresting way, if at all. It is debatable whether there even in principle could be agreement on the basic goods or the principles approach. But if there were, what is shared is only a very superficial understanding of morality that collapses as soon as any attempt to consider the concrete implications of the principles or norms is made.

If we fail to recognize the importance of allowing space for difference and debate within the bioethics enterprise, if we fail to recognize the importance of incorporating debate and dialogue among those who share radically different moralities/ideologies, religious commitments, or cultures, we risk rendering the bioethics enterprise irrelevant to many. (Why should anyone who does not share certain premises care about what those who would impose those premises or declare them to be universal have to say?) Insofar as one wants to impose particular cultural, religious or moral premises and declare them to be universal, one should at the very least recognize that this is what one is doing and avoid claiming one respects, embraces difference, and recognize that one is calling for bioethics to be a narrow ideological pursuit. Insofar as one claims to value diversity and inquiry, then one should avoid false declarations of a shared morality. If the bioethics enterprise is to continue to address a broad range of issues through a variety of functions throughout society, it is important to allow space for religious, cultural, and moral/ideological diversity within the enterprise.

7.4 Concluding Remarks

There are those who hold a particular vision of what bioethics was historically, what it is now, and what it ought to be in the future. People often speak or write as if there is a profession called "bioethics" and there are views and normative positions that necessarily do or should follow from being a member of said profession. Currently, though, the bioethics enterprise is a large umbrella under which many different issues are addressed in a variety of settings for different purposes (e.g., resolving conflicts in individual cases or establishing public policy). The enterprise involves individuals from numerous disciplines and, therefore, we lack a clear understanding of who – if anyone – is a bioethicist. The disciplinary and functional diversity we see in bioethics as well as the trend toward sub-specialization are pragmatic, as we suggested above. Yet this heterogeneity is scuh that sometimes it is difficult to see whether it even makes sense to call the various players members of the same profession. Furthermore, we argue that insofar as the bioethics enterprise wishes to engage in rational debate, discussion and analysis, we must recognize the limits of reason and allow for dialogue rather than presume consensus or restrict the scope of inquiry in bioethics. The nature of morality and normative discourse itself is debated. We should not presume that we can build the rock that many think bioethics has or should become on such shaky ground. The four ways in which bioethics is diverse are important and efforts to minimize differences along any of these axes should be resisted. "Mary is a bioethicist" should, in fact, tell us very little about Mary.

Notes

1. We focus here on the bioethics aspect of ASBH rather than the humanities. Much could be said about the medical humanities and the relationship between bioethics and the humanities, but we do not engage that discussion here. Our focus is on bioethics and we demonstrate that

7 The "s" in Bioethics: Past, Present and Future

there is significant diversity among individuals engaged in the bioethics enterprise. Arguably, ASBH is the most public and obvious embodiment of the bioethics enterprise in the United States. There are other major bioethics associations that similarly reflect the pluralism we discuss here, such as the Canadian Bioethics Society and the International Association of Bioethics.

2. We hesitate to refer to bioethics as a "field" or "discipline" because there remains controversy regarding the status of bioethics. See, for example, Kopelman (1998, 2006). As a result, we use the term enterprise. We refer here to "bioethicists" throughout the chapter even though we recognize that there is not a clear notion right now of whether anyone IS a bioethicist and, if so, what makes them so. We use the term loosely to refer to people who identify themselves as bioethicists or whom others identify as bioethicists.

3. Lisa Rasmussen has examined the question of whether bioethics has a core purpose or goal, or whether bioethicists have inherent moral obligations as bioethicists, particularly with regard to clinical consultation and consultation for corporations (Rasmussen, 2005). She has critically evaluated a number of attempts to specify a robust core that defines bioethics.

4. Our focus here is on the education of individuals who will be bioethicists rather than on teaching bioethics to medical, nursing, philosophy, and other students. For further discussion of bioethics education in general, see volume 27, number 4 of *The Journal of Medicine and Philosophy*. For an examination of ethics education in medical schools, see DuBois and Burkemper (2002). For an examination of ethics education in nursing schools, see Burkemper et al. (2007).

5. We do not mean to suggest that there is agreement about whether there are particular contributions from different disciplines individuals must master to be bioethicists.

6. One of us (ASI) has seen this comment made in regard to physicians on a bioethics listserv.

7. Our focus here is on ethics and not the medical humanities. We remind readers that in discussing only bioethics in the remainder of this paper when considering ASBH we have not forgotten about the medical humanities; this chapter is about bioethics, however, and we will not consider further the "H" in ASBH.

8. If one looks at the National Commission members and reads the roster of those who contributed to the National Commission's work, one finds many names that today are associated with bioethics, including Albert Jonsen, Patricia King, Karen Lebacz, Tom Beauchamp, and H.Tristram Engelhardt, Jr.

9. In examining the permissibility of for-profit bioethics consultation in the private sector, Lisa Rasmussen considers the objection that for-profit consultation "forsakes the purpose" of bioethics. She asks, on what basis might one claim a definitive set of goals for bioethics (whereby one can then define its purpose), and then considers arguments based on the history of bioethics. The problem with such arguments, notes Rasmussen, is that the history of bioethics lends itself to diverse interpretations (see Riech, 1994; Stevens, 2000; Jonsen, 1998). The bioethics histories locate its inception at different times, as a result of different social forces, and thus, in the end, beget different goals. Moreover, the diverse groups that first entered into bioethics had different reasons and aims for doing so. She therefore concludes that "it is implausible that a successful argument can be made for a strong essentialism about the goals of bioethics based on the history of the field" (Rasmussen, 2005, p. 96). This argument further illustrates the significance of the diversity within bioethics.

10. The regulations that mandate the existence of an IRB or REC in the United States, also sometimes called by other names, such as Human Studies Committee, do not require that the committee include a bioethicists, but many bioethicists have served on an IRB. See 45CFR46.107 for IRB membership requirements.

11. This is not to say that there never will be an issue on which people engaged in bioethics share a conclusion. It is a cautionary note that one should not assume concurrence or that disagreement is irrelevant and therefore dismissible. Some have suggested, for example, that among bioethicists there is agreement that there is an obligation to ensure universal access to health care. Yet this simply is not so.

12. Marshall has suggested that ASBH must take positions and defend justice because it should be "morally relevant" that she is a member of ASBH (2007). This seems to reflect a view

of ASBH as something other than the type of organization it was created as and its defined purposes. It makes belonging to ASBH sound more like belonging to a religious organization or a political party.

13. We find it curious that he was so sure our membership does not reflect the representation of what we believe he would call "sexual minorities" since to the best of our knowledge ASBH membership forms do not ask about one's sexual orientation and it's not at all clear how one necessarily would determine this in the context of a professional meeting.

14. The first project of ASBH's Status of the Field Commission was to survey graduate programs that trained and educated students to work in bioethics or medical humanities. The survey speaks to the significance of the disciplinary and educational diversity found within the bioethics enterprise. In all, the survey included 47 institutions and 108 programs, consisting of: 68 MA, 19 PhD, 13 fellowship, 11 certificate, and 2 "other." Among the bioethics/humanities graduate training programs, 11 different institutional homes were identified; most commonly identified were the school/college of medicine, followed by the college of arts and sciences. The survey also documents a wide array of disciplinary homes (11 specific homes recorded): most notably, philosophy, medicine, and "interdisciplinary" or "multidisciplinary." Finally, the survey underscores the diversity of the background of faculty: (aggregated for all program types) 20% philosophy, 15% medicine, 13% law, 12% theology or religious studies, 10% nursing, 6% history, 5% behavorial/mental health, 4% sociology, 4% English lit, 4% pharmacology, 2% social work, 5% "other." Overall, there was an estimated 182% increase in programs since 1990 (see Ausilio and Rothenberg, 2004).

References

Alora, A.T., and J.M. Lumitao. eds. 2001. *Beyond a Western bioethics: Voices from the developing world*. Washington, DC: Georgetown University Press.

Antommaria, A.H.M. 2004. A Gower maneuver: The American Society for Bioethics and Humanities' resolution of the "taking stands" debate. *American Journal of Bioethics* 4(1): W24–W27.

Ausilio, M.P., and L.S. Rothenberg. 2004. Bioethics, medical humanities, and the future of the field. *American Journal of Bioethics* 2(4): 3–9.

Beauchamp, T.L., and J.F. Childress. 2008. *Principles of biomedical ethics*. New York: Oxford.

Boyle, J. 1980. Contraception and natural family planning. *International review of natural family planning* 4: 309–315.

Boyle, J. 2004. Natural law and global bioethics. In *Natural law and the possibility of a global ethics*, ed. M.J. Cherry, 1–15. Dordrecht: Kluwer.

Boyle, J. 2006. The bioethics of global biomedicine. In *Global bioethics: The collapse of consensus*, ed. H.T. Engelhardt, Jr. 300–334. Salem, MA: Scrivener Press.

Brody, B., et al. 2002a. Bioethics consultation in the private sector. *Hastings Center Report* 32(3): 14–20.

Brody, B., et al. 2002b. The task force responds. *Hastings Center Report* 32(3): 22–23.

Buchanan, A., and D. Brock. 1990. *Deciding for others: The ethics of surrogate decision making*. New York: Cambridge University Press.

Burkemper, J., J.M. DuBois, M.A. Lavin, G.A. Meyer, and M. McSweeney. 2007. Ethics education in MSN programs: A study of national trends. *Nursing Education Perspectives* 28(1): 10–17.

Callahan, D. 1973. Bioethics as a discipline. *Hastings Center Studies* 1(1): 66–73.

Chattopadhyay, S., and A. Simon. 2008. East meets West: Cross-cultural perspective in end-of-life decision making from Indian and German viewpoints. *Medicine, Health Care and Philosophy* 11(2): 165–174.

Chen, Xiao-Yang. 2007. Defensive medicine or economically motivated corruption? A Confucian reflection on physician care in China today. *Journal of Medicine and Philosophy* 32(6): 635–648.

Chen, Xiaoyang, and Ruiping Fan. 2010. The family and harmonious medical decision making: Cherishing an appropriate Confucian moral balance. *Journal of Medicine and Philosophy* 35(5): 573–586.

Cho, M.K., S.L. Tobin, H.T. Greely, J. McCormick, A. Boyce, and D. Magnus. 2008. Strangers at the bedside: Research ethics consultation. *American Journal of Bioethics* 8(3): 4–13.

Derse, A.R. 2005. The H in ASBH. *ASBH Exchange* 8(2): 2.

DeVries, R. 2004. Businesses are buying the ethics they want. *Washington Post*, February 8, p. B2.

DeVries, R., and C. Bosk. 2004. The bioethics of business. *Hastings Center Report*, Sept–Oct, 28–32.

DuBois, J., and J. Burkemper. 2002. Ethics education in U.S. medical schools: A study of syllabi. *Academic Medicine* 77(5): 432–437.

Elliott, C. 2001. Pharma buys a conscience. *The American Prospect* 12(17): 16–20.

Engelhardt, H.T., Jr. 1996. *The foundations of bioethics*. New York: Oxford.

Engelhardt, H.T., Jr. 2002. The ordination of bioethicists as secular moral experts. *Social Philosophy and Policy* 19(2): 59–82.

Engelhardt, H.T., Jr. 2006. Public discourse and reasonable pluralism: Rethinking the requirements of neutrality. In *Handbook of bioethics and religion*, ed. D.E. Guinn, 169–194. New York: Oxford University Press.

Engelhardt, H.T., Jr., and L.M. Rasmussen. eds. 2002. *Bioethics and moral content: National traditions of health care morality*. Dordrecht: Kluwer.

Fan, Ruiping. 2004. Truth telling in medicine: The Confucian view. *Journal of Medicine and Philosophy* 29(2): 179–193.

Fan, Ruiping. 2007. Which care? Whose responsibility? And why family? A Confucian account of long-term care for the elderly. *Journal of Medicine and Philosophy* 32(5): 495–517.

Finnis, J., J. Boyle, and G. Grisez. 1987. practical principles, moral truth, and ultimate ends. *American Journal of Jurisprudence* 32: 99.

Gbadegesin, S. 1993. Bioethics and culture: An African perspective. *Bioethics* 7(2–3): 257–262.

Hoshino, K. 1997. *Japanese and Western bioethics: Studies in moral diversity*. Dordrecht: Kluwer.

Iltis, A. 2005. Bioethical expertise in health care organizations. In *Ethics expertise: History, contemporary perspectives, and applications*, ed. L.M. Rasmussen, 259–267. Dordrecht: Springer.

Iltis, A.S. 2006. Look who's talking: Interdisciplinarity and the implications for bioethics education. *Journal of Medicine and Philosophy* 31(6): 629–641.

Institute of Medicine. 2006. *Organ donation: Opportunities for action*. Washington, DC: National Academy of Sciences.

Jonsen, A.R. 1998. *The birth of bioethics*. New York: Oxford University Press.

Jonsen, A.R. 2000. *A short history of medical ethics*. New York: Oxford University Press.

Jonsen, A.R. 2001. Social responsibilities of bioethics. *Journal of Urban Health* 78(1): 21–28.

Jonsen, A.R., and R.C. Fox. 1996. Bioethics, our crowd, and ideology. *Hastings Center Report* 26(6): 3–8.

Kelly, G.S.J. 1949–1954. *Medico-moral problems, Parts I–V*. St. Louis: The Catholic Hospital Association of the United States and Canada.

Kelly, G.S.J. 1958. *Medico-Moral Problems*. St. Louis: The Catholic Hospital Association of the United States and Canada.

Kipnis, K. 1997. Confessions of an expert ethics witness. *Journal of Medicine and Philosophy* 22: 325–343.

Kopelman, L.M. 1998. Bioethics and humanities: What makes us one field? *Journal of Medicine and Philosophy* 23(4): 356–368.

Kopelman, L.M. 2006. Bioethics as a second-order discipline: Who is not a bioethicist? *Journal of Medicine and Philosophy* 31(6): 601–628.

Kosmin, B., E. Mayer, and A. Keysar. 2001. *American religious identification survey*.New York: The Graduate Center of the City University of New York.

Kumar, N.K. 2006. Bioethics activities in India. *Eastern Mediterranean Health Journal* 12(Supplement 1): S56–S65.

Lantos, J.D. 2005. Commentary on "a draft model aggregated code for bioethicsts". *American Journal of Bioethics* 5(5): 45–46.

LaRochelle, S.A., and C.T. Fink. 1947. *Handbook of medical ethics for nurses, physicians and priests* (trans: Poupore, M.E., A. Carter, and R.M.H. Power). Westminster, MD: The Newman Bookshop.

Li, En-Chang, and Chun-Feng Wen. 2010. Should the Confucian family-determination model be rejected? A case study. *Journal of Medicine and Philosophy* 35(5): 587–599.

Loewy, E. 2007. Healthcare systems and motivation. *Medscape General Medicine* 9(1): 41.

Marshall, M.F. 2007. ASBH and moral tolerance. In *The ethics of bioethics: Mapping the moral landscape*, eds. Lisa Eckenwiler and Felicia Cohn, 134–143. Baltimore, MD: Johns Hopkins University Press.

McCullough, L.B. 1998. *John Gregory and the invention of professional medical ethics and the profession of medicine*. Dordrecht: Kluwer.

de Melo-Martín, I., L.I. Palmer, and J.J. Fins. 2007. Viewpoint: Developing a research ethics consultation service to foster responsive and responsible clinical research. *Academic Medicine* 82(9): 900–904.

Miller, F.G. 2005. The case for a code of ethics for bioethicists: Some reasons for skepticism. *American Journal of Bioethics* 5(5): 50–52.

Montgomery, K. 2002, March. Interdisciplinarity in bioethics and humanities. *ASBH Exchange* 2:10.

National Bioethics Advisory Commission. 1998. *Research involving persons with mental disorders that may affect decision making capacity*. Rockville, MD: NBAC.

National Bioethics Advisory Commission. 1999. *Ethical issues in human stem cell research*. Rockville, MD: NBAC.

Parker, L. 2007. Bioethics as activism. In *The ethics of bioethics: Mapping the moral landscape*, eds. A. Eckenwiler and Felicia G. Cohn, 145–157. Baltimore, MD: Johns Hopkins University Press.

Parsi, K., and K. Geraghty. 2004. The bioethicist as public intellectual. *The American Journal of Bioethics* 4(1): 17–23.

Pellegrino, E.D. 2000. Bioethics at century's turn: Can normative ethics be retrieved? *Journal of Medicine and Philosophy* 25(6): 655–675.

Pessini, L., C.D.P. de Barchifontaine, and F. Lolas Stepke. eds. 2010. *Ibero-American bioethics*. Dordrecht: Springer.

President's Council on Bioethics. 2004. *Monitoring stem cell research*. Washington, DC: PCB.

Rasmussen, L. 2005. The ethics and aesthetics of bioethics consultation. *HEC Forum* 17(2): 94–121.

Rasmussen, L. ed. 2005. *Ethics expertise: History, contemporary perspectives and applications*. Dordrecht: Springer.

Reich, W.T. 1993. How bioethics got its name. *Hastings Center Report* 23(6): S6–S7.

Reich, W.T. 1995. The word "bioethics": Its birth and the legacies of those who shaped its meaning. *Kennedy Institute of Ethics Journal* 5: 319–335.

Reich, W.T. 1999. The "wider view": André Hellegers's passionate, integrating intellect and the creation of bioethics. *Kennedy Institute of Ethics Journal* 9(1): 25–51.

Rothman, D.J. 1991. *Strangers at the bedside: A history of how law and bioethics transformed medical decision making*. New York: Basic Books.

Stevens, T. 2000. *Bioethics in America: Origins and cultural politics*. Baltimore: Johns Hopkins University Press.

Tangwa, G.B. 2000. The traditional African perception of a person: Some implications for bioethics. *The Hastings Center Report* 30: 39–43.

Tao Lai Po-Wah, J. ed. 2002. *Cross-cultural perspectives on the (Im)possibility of global bioethics*. Dordrecht: Kluwer.

UNESCO. 2005. Universal declaration on bioethics and human rights. Available at: http://www.unesco.org/new/en/social-and-human-sciences/themes/bioethics/bioethics-and-human-rights/. Accessed 21 Jan 2011.

Veatch, R. 2005. *Death, dying, and the biological revolution: Our last quest for responsibility.* New Haven, CT and London: Yale University Press.

Wang, M., and X. Wang. 2010. Organ donation by capital prisoners in China: Reflections in Confucian ethics. *Journal of Medicine and Philosophy* 35(2): 197–212.

Wolpe, P.R. 2007. A rocky mountain challenge. *ASBH Exchange* 10(1): 2.

Youngner, S., and R. Arnold. 2002. Who will watch the watchers? *Hastings Center Report* May–June: 21–22.

Chapter 8
Why Clinical Bioethics So Rarely Gives Morally Normative Guidance

H. Tristram Engelhardt, Jr.

8.1 Bioethics as a Complex Social Phenomenon

What is bioethics, that it could so rapidly establish its academic and social status in the United States and then almost as quickly be accepted across the world? The emergence of bioethics is one of the remarkable cultural occurrences of the 20th century. Within seven years after the founding of the Kennedy Institute where the term bioethics was re-coined,[1] the first edition of *The Encyclopedia of Bioethics* had been published.[2] By the 1980s, bioethics was widely recognized as both an academic and a quasi-clinical profession, such that it is now impossible to imagine American medical schools without courses in bioethics or without ethicists serving on their Institutional Review Boards. For that matter, it is hard to imagine academically affiliated hospitals without ethics committees or bioethics consultants (health care ethics consultants). Bioethics and the presence of clinical ethicists have become part of the everyday lifeworld of not just the United States, but of much of Europe, the Pacific Rim, and elsewhere. In some forty years, bioethics has come to command a significant cultural and social authority.

There are a number of accounts of the emergence of bioethics and bioethicists.[3] These accounts have generally failed adequately to recognize that bioethicists gained their importance through serving as secular substitutes for theologians and priests (Engelhardt, 2002), as well as by functioning as lay replacements for authoritative physicians, not primarily by providing normative guidance but by offering a cluster of other services. There has generally been a failure sufficiently to acknowledge the character of the socio-cultural forces that created a field that compasses a bundle of services, only a small portion of which involve giving frankly morally normative direction. This essay begins by exploring the moral and cultural vacuum that spurred the rapid emergence of bioethics. This history of the genesis of bioethics is tied to the underappreciated heterogeneity of the field itself, as well as to the circumstance that bioethics is as much a social movement as an intellectual

---

H.T. Engelhardt, Jr. (✉)
Department of Philosophy, MS-14, Rice University, Houston, TX 77005, USA
e-mail: htengelh@rice.edu

H.T. Engelhardt, Jr. (ed.), *Bioethics Critically Reconsidered*,
Philosophy and Medicine 100, DOI 10.1007/978-94-007-2244-6_8,
© Springer Science+Business Media B.V. 2012

discipline. All of this has led to the circumstance that bioethics succeeds without primarily giving normative guidance. Although bioethics appears to promise morally normative direction, for the most part it offers other services. The social forces that engendered and sustained bioethics have produced a complex social movement able to sustain itself by bringing under its rubric a number of important academic and clinical services.

A number of changes in 20th-century American culture created a moral vacuum, which a field such as bioethics could fill. The late 1950s and especially the 1960s were marked by a felt need to find moral orientation and to recapture a sense of what it means to live a humane life. The secularization of American society and similar processes of secularization in Western Europe marginalized traditional religious moral experts. For those committed to this secularization, it was not enough to disestablish religious moral experts, but it was also necessary to create new moral experts to sustain the emerging secularism.[4] This secularization occurred nearly in tandem with the recasting of the status of the medical profession. A series of anti-trust court cases had the effect of disestablishing medicine as a guild, thus indirectly bringing into question the traditional role of physicians as judges of the moral norms that should direct their own profession. The status of physicians as guides for the proper conduct of medicine was undercut. In part, this was the case because the anti-elitist sentiments of the time made it plausible that the moral norms of physicians should be open to critical assessment and approval by all. The secularization of society and the marginalization of medical ethics were expressions of wide-reaching cultural developments that combined commitments to secularity and to general social-democratic values with anti-elitist sentiments.

These forces opened up a moral, social, and economic niche into which bioethicists, including clinical ethicists, were able to enter and successfully market their services. Bioethics, shaped as it was by a diversity of social and cultural forces, came to constitute a complex amalgam of disciplines and services. These congealed into a hybrid field, which has succeeded in selling different services to different customers under different circumstances. In the face of moral pluralism and despite bioethics' incapacity to define a uniting moral goal or a cardinal social role for itself as a "profession", bioethicists continued nevertheless to be sought out, especially in hospitals, for "ethical guidance" and "ethics consultations". In all of this, clinical ethicists rarely give frankly morally normative advice, although academic bioethicists have been defenders of a large plurality of morally normative bioethical positions. In the clinic and in the provision of clinical ethics consultations, bioethicists have for the most part provided legal advice bearing on advance directives, surrogate decision-making, end-of-life decisions, informed consent, etc., all under the color of giving ethical guidance. While clinical ethicists (health care ethics consultants) offer to provide *obiter dicta* regarding moral analyses and moral theories, and even though they may express their own moral convictions, the core services they provide are quasi-legal. Clinical ethicists are in general experts regarding the ethos established at law and through public policy. As quasi-lawyers, clinical ethicists have also succeeded as "clinical ethicists for hire". Just as one might go shopping for a lawyer who will be sympathetic to one's case, ethicists have been hired in the support of various political and ideological agendas. They have been

8 Why Clinical Bioethics So Rarely Gives Morally Normative Guidance

hired as well by for-profit corporations, who understand that bioethicists can serve as advocates of a favorable interpretation of how those corporations can operate in the grey zones of the ethos established at law. The role of bioethicists as ethicists for hire reflects the capacity of ethicists to negotiate on behalf of their clients within the legally established ethos. Within this established ethos, clinical ethicists (health care ethics consultants) serve as legal liability risk managers and dispute mediators (as do lawyers), as well as ethics experts (e.g., concerning moral theory and different possible moral positions). Bioethics has proven much more complex than any had anticipated at its inception.

8.2 The Cultural-Moral Vacuum into which Bioethics Stepped

Given the widespread and now firmly established contemporary secularization of American public culture, it is difficult to appreciate what American culture was like before the changes in culture and law that occurred during the last century. It is difficult fully to appreciate that America was de facto and de jure a Christian country until the mid-20th century,[5] such that with the exception of southern Louisiana and a few other jurisdictions, Protestantism was culturally and legally equivalent to an established religion for the United States. As the Supreme Court of the United States in 1892 acknowledged, affirming an early 19th-century Pennsylvania state court ruling (*Updegraph v. The Commonwealth*, 11 S & R 394 [1824] at 400), "Christianity, general Christianity, is, and always has been, a part of the common law of Pennsylvania;...not Christianity with an established church, and tithes, and spiritual courts; but Christianity with liberty of conscience to all men."[6] Granted, Christianity was not established in the sense of generally having its ministers' salaries paid by the state, yet in a taken-for-granted fashion Christianity shaped the fabric of much of American law and public policy. As late as the 1930s, the Supreme Court opined, "We are a Christian people, according to one another the equal right of religious freedom, and acknowledging with reverence the duty of obedience to the will of God."[7] Even in mid-20th century, the Supreme Court could affirm, "We are a religious people whose institutions presuppose a Supreme Being."[8] For most of American history, most of the citizens of the United States thought of themselves as Christian, and of the United States as a Christian polity, indeed a Protestant polity.[9] This establishment expressed itself in public prayer and Protestantly-oriented Bible readings in public schools, as well as in an attempt to suppress Orthodox Christianity in Alaska by the Presbyterian minister Sheldon Jackson (1834–1909), when he served as Alaska's first General Agent for Education (Oleksa, 1992).

In mid-20th century, American Supreme Court decisions effected a substantive secularization of American law and public policy. These decisions reached to holding that the State of Maryland could not require state officeholders, including notary publics, to declare a belief in God.[10] They forbade public schools from requiring and eventually even allowing the recitation of the Lord's Prayer and the reading of Bible passages in regular school programs.[11] Other Supreme Court rulings disestablished Christian morality, most significantly forbidding the legal prohibition of abortion under most circumstances.[12] These court holdings were part of a profound cultural struggle aimed at redefining the public forum and transforming public discourse so as to render it canonically secular. This sea change in the dominant public culture continues in attempts to remove all public remnants of America's Christian and more generally religious past.[13] As to the latter, one might think of suits against the American motto "In God we trust."[14] The changes in American culture, which created the background for the emergence of bioethics, brought to America a secularization that had emerged in Europe, especially in France, with the Enlightenment, the French Revolution, and Napoleon. As Peter Gay observes, the Enlightenment involved a commitment to establishing a pagan moral and metaphysical vision. "The philosophes' experience [involved] a dialectical struggle for autonomy, an attempt to assimilate the two pasts they had inherited—Christian and pagan—to pit them against one another and thus to secure their independence. The Enlightenment may be summed up in two words: criticism and power. . . . I see the philosophes' rebellion succeeding in both of its aims: theirs was a paganism directed against their Christian inheritance and dependent upon the paganism of classical antiquity but it was also a *modern* paganism, emancipated from classical thought as much as from Christian dogma" (Gay, 1995, p. xi). It is the case that Jerome Schneewind has argued that the Enlightenment was not directed so much against religion, as against clericism (Schneewind, 1998). His account underestimates the animus of the Enlightenment against traditional Christianity, an animus expressed in views such as those of David Hume, who used Christianity as a term of derision, as exemplified in his letter to the Rev. Hugh Blair (April 6, 1765), where Hume speaks of the English as "relapsing fast into the deepest stupidity, Christianity, and ignorance" (Hume, 1932, vol. 1, p. 498). A culture emerged that was anti-Christian and on its way to being post-Christian.

All of this occurred in the shadow of the religious wars of the 16th and 17th centuries, as Western Christendom with its marriage of faith and reason fell stepwise into ruins. In 1648 a crucial development occurred as the very sense of Christendom fell into ruins in the West.[15] Many of the intellectual class became openly alienated from Christianity and Christian theology, giving issue to critical anti-Christian sentiments that lay at the foundation of the Enlightenment. The result was that the Enlightenment, despite its putative commitment to peace, culminated in the French Revolution, which involved a violent, vigorous, and bloody rejection of Western Europe's Christian commitments.[16] By the late 18th century and early 19th century, a secular cultural perspective had gained salience in Western Europe that was robustly laicist. The French Revolution, the Josephism of Joseph II, Emperor of Austria and the Holy Roman Empire (1765–1790), the imposition by Napoleon of a new secular legal framework, and the disestablishment of Christianity in many areas of society, especially through the Secularization of 1803 (authorized on August 24, 1802), created a public culture marked by a widespread abandonment of faith among the intelligentsia. A post-Christian Europe had begun to emerge. In particular, the impact of the German secularization should not be underappreciated.[17] By the mid-20th century, these cultural developments had produced a secularist if not laicist movement that aimed at removing the remaining legal, institutional, and public policy marks of Christendom. The result was that for many in late 20th-century America it no longer appeared legitimate or even possible to draw moral direction for medicine and the biomedical sciences from moral-theological sources. A new secular source of medical-ethical guidance had to be found, indeed created.[18] Bioethics emerged in a normative or dominant culture which had become secular.

The secularization of expectations regarding the sources for moral guidance in the use of medicine and the biomedical sciences occurred as there was also an alteration in the social and legal standing of medicine, physicians, and medical ethics. The medical profession ceased to function as a quasi-guild and was reduced to a trade (Krause, 1996). Physicians had from Hippocratic times considered the preservation of the profession's standing as integral to its members' proper functioning as physicians. The self-identity of physicians had for millennia been strengthened by often elitist concerns with professional ethics and self-governance.[19] Initially, the emergence of modern scientific medicine had supported the medical professional's claim to be a natural elite grounded in knowledge and expertise. As the 20th century emerged, physicians were regulating themselves as a quasi-guild, within which medical-moral commitments, economic concerns, and a sense of professional identity were intertwined in their professional ethics. The traditional authority of physicians was in part undermined when the Supreme Court defined physicians no longer as members of a quasi-guild able freely to articulate its own professional norms.[20] Medicine as a profession was also brought into question by the new field of bioethics.

Medical ethics, which had been grounded in the clinical experience and the authority of physicians, was relocated within the authority of bioethics, because bioethics was held to reflect the claims of general secular rationality. Within the emerging dominant cultural milieu, it was no longer plausible for physicians to claim special moral insights born of clinical experience into the proper practice of medicine that were not open to persons generally. If a special, clinically derived, medical professional insight into the appropriate moral norms for the practice of medicine, grounded in a clinical moral epistemology, did not exist, and if the dominant culture of the public forum was to become secular, bioethicists could supply what was wanting.[21] Absent crediting a clinical medical-moral epistemology, professional medical ethics could no longer claim a privileged sphere unto itself, thus bringing to an end centuries of medical-ethical reflections.[22] Professional ethics were then put into the hands of a group of moral experts who were independent of the medical profession. This transformation of the profession of medicine occurred along with the secularization of American society, creating a moral vacuum that enabled and accelerated the development of bioethics, which then further marginalized medical ethics and even further deflated the moral authority of physicians. Bioethics was thus warranted to judge the claims of professional medical ethics.

The impetus to articulate and develop bioethics also occurred as traditional authority figures were generally brought into question by the various civil rights movements of the 1960s and 1970s. A dialectic of suspicion was created against tradition and against those traditionally in authority, leading, *inter alia*, to the replacement of the professional standard for consent to medical treatment with the objective or reasonable-and-prudent-person standard.[23] Even elites that grounded their status in claims to scientific knowledge and technological expertise were brought into question. There was a concern to circumscribe the authority of those who were accepted as scientific and technological experts. The result was that now patients themselves, not physicians, should determine what information should prudently bear on their medical decisions. It was not just that physicians ceased to be custodians of the moral norms bearing on medicine, but they were no longer generally recognized as having insight into what should direct medical decisions. Bioethics emerged, claiming to be able to give guidance in this circumstance, in that they claimed to be innocent of the self-interest and elitism of physicians.

Bioethicists appropriated what had been addressed by medical ethics and medical moral theology. They then re-interpreted and re-located medical-moral concerns within secular, non-professional moral reflections that they asserted were open to all. In the process, the moral standpoint that should guide medicine was rendered external to medicine and beyond the bounds of theological reflection. The experience of clinical engagement and the direction of moral theology that had previously been accepted as sources for medical moral knowledge were marginalized and deflated, just as a source of guidance for an emerging, powerful, medical-industrial complex was direly needed. In addition, the bioethics that emerged was a function of a secular, social social-democratic ethos that combined the secularization of society and of culture with a general distrust of traditional authority figures. That is, the founding of a bioethics was shaped by persons generally who were at peace in the liberal wing of the Democratic Party. All of this occurred as medicine became powerful, expensive, and provocative of moral puzzles.

8.3 The Emergence of Salient Moral and Metaphysical Pluralism

Underlying the processes of secularization of law and public policy and of the deprofessionalization of medicine were even more profound changes in the background culture that finally brought bioethics itself into question. Moral rationality, rather than achieving a unity of vision, was ever more recognized as having fragmented into a plurality of moralities and accounts of the morally reasonable. This fragmentation was always in principle present, because in order to establish any particular content-rich moral vision as canonical, one must already possess common and canonical basic premises and rules of evidence so as to be able authoritatively to rank or order cardinal values as well as right-making conditions as a first step to establishing a canonical morality. This plurality of moral and bioethical visions remains, even if the building blocks that frame the competing moralities and bioethics are in general very similar. The result is that it may even be the case that all or almost all moralities with their bioethics affirm similar values and similar right-making conditions. However, different rankings of those values and right-making conditions frame divergent moralities. It may be the case that all moralities are concerned with when it is appropriate to kill humans, have sex, reproduce, tell lies, and take the property of others, but the plurality of moralities persists, because moralities differ at least in terms of when it is obligatory, licit, or forbidden to kill humans, have sex, reproduce, tell lies, and take others' property.

Even if some moral propositions appear to be universally accepted, on closer examination one encounters moral pluralism. For example, although the proposition, "It is immoral to cause great suffering to the innocent for no purpose whatsoever," may appear to be accepted by all, it shatters into a pluralism of views as soon as one examines what is packed into "innocent" and "no purpose". *Ceteris paribus*, suffering considered in isolation may be disvalued, but the actual evaluation of the justification for causing suffering depends on particular framing normative and metaphysical views. Such supposedly universal moral propositions do not establish a particular, canonical, concrete view of how to act benevolently or rightly. Such propositions gain content only when one adds when it is obligatory, forbidden, or licit to cause suffering for particular purposes. An actual content-full morality *cum* actual content-full bioethics requires a cluster of settled moral judgments grounded in particular rankings of values and goods, as well as of right- and wrong-making conditions. The further difficulty is that a particular ranking cannot be established as morally normative without first conceding particular moral premises and rules of evidence. Moralities are concerned with the good and the right, with beneficence and benevolence, but there are profound disagreements as to what concretely constitutes the good and the right. The plurality of moral visions and bioethics is further compounded by a plurality of possible rational reconstructions of such concrete moral lifeworlds. Moral philosophical accounts of moralities can themselves further shape those moralities and increase moral pluralism.

There is no easy way out of this state of affairs. John Rawls (1921–2002), for example, came to appreciate the challenge of moral pluralism during his retreat from an account of morality as rationally normative to an account of justice, as would seem reasonable within a particular political viewpoint. His retreat reflected the retreat of many academic moral theorists from the view that one can secure a foundationally grounded canonical moral account. Although *A Theory of Justice* might initially have appeared as able to be read in a quasi-Kantian mode as an exegesis of the claims of moral rationality as such, so as to govern all persons,[24] the later Rawls takes pains to place his account of justice and fairness within the weaker claims of a view of the politically reasonable. Rawls recognizes the limits of moral rationality and abandons the project of establishing a particular canonical account of moral rationality when he concedes that his account is not grounded in reason itself or in a canonical morality (Rawls, 1985), but in a freestanding view (Rawls, 1993, p. 25). He then attempted to advance a socio-democratic political agenda he wished to embrace as framing a reasonable account by appealing to quasi-empirical considerations regarding the likelihood of this view persisting over generations (Rawls, 1993, p. 15). The mature account given by Rawls is thus embedded within the particular political paradigm of a social democratic constitutional regime (Rawls, 1993). However, history has always been hard to predict, and the usual office of the philosopher has been to understand what is, not what will be. The owl of Minerva spreads her wings only at dusk.[25] At the very least, it is very difficult to foresee what social arrangements will be stable over time.

The retreat of Rawls and the recasting of his moral-philosophical project are a function of his confrontation with moral pluralism and the limits of moral-philosophical rationality. The medieval Western cultural-philosophical synthesis with its strong faith in reason (and not just faith in faith), the modernity that took shape after the Reformation and the Renaissance, as well as the Enlightenment that continued the peculiar Western Christian commitment to moral-philosophical rationality, aspired to disclosing a rationally principled basis for surmounting the challenges of moral pluralism. These cultural perspectives predicted progress towards unanimity and consensus, but these hopes shipwrecked on intractable moral pluralism. Moral pluralism has been both recognized and denied since the ancient world; it persists and remains central. Indeed, the failure of foundationalism and the confrontation with post-modernity indicate that moral pluralism and its challenges have an intractable character. That is, there is no way free of moral pluralism if one cannot identify the correct basic moral premises or rules of evidence. This, however, is not possible unless there is an incorrigible union of the knower with the known. The failure of the modern philosophical project of establishing a universal, rationally grounded, univocal appreciation of truth, including moral truth and bioethics, is tied to the recognition that secular, discursive rationality does not possess a *nous*, the possibility for achieving such a unity.

8.4 The Immanent Domestication of the Transcendent: The Deeper Roots of Post-modernity and Moral Pluralism

Because philosophical reflection cannot bring the knower into union with the known as the known is apart from the knower, one is left with impressions of a reality that is always distinct from those impressions. The result is that within the compass of discursive reflection, one never knows the known directly, as it is in itself. Instead, one only has an experience of the known as it is for the knower. Without reference to a God's-eye perspective in terms of which one can refer to a canonical historical reality, there is no longer a canonical appreciation of humanity and human rights. As Giannni Vattimo recognizes, much of contemporary moral and cultural disorientation is a consequence of a loss of God as an ultimate point of reference: "We might begin this discussion of humanism by paraphrasing a joke that went around some time ago, pointing out that in the contemporary world 'God is dead, but man isn't doing so well himself.' Although in one sense this is just a joke, it is also something more than that, inasmuch as it captures and conveys the difference between contemporary atheism and the more classic kind of atheism represented by Feuerbach. ... the death of God, which is at once the culmination and conclusion of metaphysics, is also the crisis of humanism" (Vattimo, 1991, pp. 31, 32–33). The entire view of reality is recast.

8 Why Clinical Bioethics So Rarely Gives Morally Normative Guidance 159

This state of affairs leads to recognition of the intractability of moral pluralism and the collapse of classical metaphysics as the study of being as it is in itself, and its replacement with the examination of being insofar as it is for us, that is, insofar as it offers itself within appearance, within language, within culture. Those perspectives are multiple. Given this isolation of the knower from reality as it is in itself, one is led to hold that all that being can be in itself is what being can be for us, with the result that the constitution and character of being are determined by our experience and our reflection. This state of affairs began once more to be widely recognized from the end of the Enlightenment to the beginning of post-modernity. Both David Hume (1711–1776) and Immanuel Kant (1724–1804) in different ways appreciated that within secular reflection we are inescapably imprisoned within the horizon of finite, immanent, sensible experience, the coherence and intelligibility of which experience we ourselves shape, indeed constitute.

Reality along with morality and bioethics came to be accepted as nested within a historically and socially conditioned human rationality. The only final standpoint available to secular reflection was seen to be located within the horizon of the finite and the immanent. As a consequence, this finite standpoint took on an absolute character. That is, reality as it is in itself, that which we can hold to be the truth of reality and morality apart from us, became only what reality and morality are, or at least can be, for us. This conclusion is the substance of G.W.F. Hegel's (1770–1831) view that, given our imprisonment within immanence, that imprisonment may in principle be ignored with the result that immanent rationality can be regarded as absolute. Hegel in this fashion immanentizes the transcendent. The transcendent is rendered into mere moments or elements of our appreciation of what it is for reality to be given to us within the sphere of the finite, immanent, and sensible. There is no further standpoint except our standpoint: being as it is in itself becomes historically and culturally conditioned. All that one could every say about being is taken to be a cultural product.

Once the step to immanentize the transcendent is taken, being must be considered to change as our thought about being changes. As a consequence, as categories change, the world itself must be seen to change. As Hegel appreciated, "All cultural change reduces itself to a difference of categories. All revolutions, whether in the sciences or world history, occur merely because spirit has changed its categories in order to understand and examine what belongs to it, in order to possess and grasp itself in a truer, deeper, more intimate and unified manner" (Hegel, 1970, § 246 Zusatz, vol. 1, p. 202). What this means, as Louis Dupré acknowledges, is that "Not merely our thinking about the real changes: reality itself changes as we think about it differently. ... [it is not] possible to capture that changing reality in an ahistorical system" (Dupré, 1993, p. 6). This step in appreciating the predicament of discursive, rational philosophical reflection and thought leads to the recognition that accounts of being, and therefore the ways being can be for us, are irreducibly multiple. This state of affairs constitutes the core of the challenge of post-modernity. If one characterizes the modern project as René Descartes' (1596–1650), Baruch Spinoza's (1632–1677), Gottfried Leibniz's (1646–1716), and Immanuel Kant's (1724–1804) attempt through discursive reflection to know reality ahistorically, then post-modernity, among other things, involved the recognition that within the horizon of the immanent and the finite being can only be for us as socio-historically conditioned and therefore framed within diverse narratives. The modern moral philosophical project shatters into a plurality of accounts, narratives, moral visions, and bioethics, all set within a plurality of self-enclosed hermeneutic circles which in their diversity constitute post-modernity. Each offers an alternative experience of reality, morality, and bioethics.

Absent possessing a nous, there is no escape. No definitive choice can be made among the alternative narratives, moralities, and bioethics on the basis of unbiased, discursively rational, principles. The frank acknowledgement of the salience of moral and metaphysical pluralism constitutes post-modernity. It marks the failure of the Enlightenment to secure a universal moral account and/or narrative concerning the significance or reality. As Lyotard puts it, "In contemporary society and culture—postindustrial society, postmodern culture ... The grand narrative has lost its credibility, regardless of what mode of unification it uses, regardless of whether it is a speculative narrative or a narrative of emancipation" (Lyotard, 1984, p. 37). The moral and bioethical pluralism, which as a result becomes salient, is grounded in the gulf between the knower and the known, between knower and reality. This state of affairs leads to two cardinal epistemological difficulties that sustain the salient moral and metaphysical pluralism that challenges the Enlightenment aspirations of bioethics and characterizes the post-modern cultural landscape within which clinical bioethics is practiced. First, moral claims are embedded within the horizon of the immanent and the finite, so that secular morality and bioethics can give no ultimate orientation or direction.[26] Second, those making moral claims find themselves always nested in one among a number of competing accounts of the immanent and finite, one among numerous disparate moral and metaphysical visions, accounts, or narratives, among which it is impossible to choose conclusively on the basis of sound rational argument without begging the question, arguing in a circle, or engaging in infinite regress. The result is a diversity of bioethics nested within the plurality of competing alternative moral visions, moral narratives, philosophical accounts, and hermeneutic circles.

8.5 Inventing a New Discipline: Bioethics as a Medieval/Enlightenment Moral Project Now Located in the Midst of Post-modernity

Remarkably, this intellectual and cultural state of affairs did not abort the birth of bioethics. Bioethics emerged in the face of moral pluralism and despite the emergence of post-modernity. Most of the pioneers of bioethics had a strong faith in reason born of the High Middle Ages and preserved in Roman Catholicism, or, in the case of some, a robust Enlightenment faith in philosophical rationality, autonomy, and progress. Sargent Shriver and André Hellegers through the Center for Bioethics at the Kennedy Institute could in good conscience advance the promises of bioethics

8 Why Clinical Bioethics So Rarely Gives Morally Normative Guidance

without acknowledging the threat of moral pluralism, because as Roman Catholics they likely did not see or fully appreciate the threat. Roman Catholic theological reflection in the 1970s, which formed the background commitments of the Kennedy Institute, maintained a strong confidence in the capacities of moral philosophy. The field of bioethics could in this context emerge with the bold promise that it was a discipline that required faith neither in a faith nor in the particular clinical experience of physicians, but only rational philosophical reflection. The purveyors of bioethics could in good conscience promise to give medicine and the biomedical sciences the moral direction they sought in terms open to all. Bioethics as it emerged reflected the Roman Catholic faith in reason born of the High Middle Ages, as well as that faith as it had been recast during the Enlightenment.

Against this background, the promises of bioethics appeared real and substantive. In particular, bioethics, it seemed, could provide moral orientation. The view was that bioethics could

(1) disclose through rational reflection the canonical content of medical morality,
(2) establish the moral authority of appropriate health care policy and law by showing these to be grounded in a rationally justified, underlying morality that could give canonical moral content to bioethics, and thus also
(3) show that because all persons are members of one moral community bound by one common morality, articulable in a single, canonical, eventually global bioethics, so that all persons as bound by one moral rationality should see themselves as members of one moral community, thus justifying the universalist aspirations of bioethics.

Morality and rationality were interconnected in a way that promised the ability of secular morality to speak universally concerning human rights and the claims of bioethics, thus establishing a community that should set aside the particularistic moral claims of religions and regional cultures in favor of a global morality and bioethics.

The result was that moral pluralism and the emergence of post-modernity did not abort bioethics. Shriver and Hellegers' creation of bioethics as a response to the moral vacuum engendered by the secularization of society and the marginalization of medical ethics had an audience that still affirmed a robust faith in the promises of moral-philosophical reasoning. Given their Roman Catholic faith in reason's capacity to disclose the general canons of moral probity, and/or given their other background Enlightenment cultural and moral assumptions, the founders of bioethics could be confident that their new field, bioethics, would fill the developing moral vacuum. They recognized society's needs for intellectual orientation as well as for concrete guidance about how properly to act in actual clinical and policy contexts, and thus they felt assured regarding the acceptance bioethics would receive (Engelhardt, 2000, chapter 1). As a consequence, they could with confidence create an institutional locus that brought together academicians as theoreticians, while also developing one-week total-immersion courses to train practitioners for the new field. The creation of bioethics as an academically respectable undertaking conveyed a legitimacy to the new bioethics practitioners who offered their services in hospitals and elsewhere. The need to train clinical practitioners legitimated the existence of the bioethics of the academy. Bioethics as we now know it came into existence through supporting its theoretical and applied undertakings.

Bioethics accomplished what the medical humanities, concerns with literature and medicine, and undertakings in the history of medicine failed to do. Bioethics achieved cultural centrality and cardinal cultural normativity. In particular, bioethics succeeded in channeling numerous, often amorphous, cultural concerns into the establishment of a class of publicly recognized moral experts, who came to be accepted as norm-setters. In the 1950s, 60 s, and 70 s, there had been a broad concern with what was perceived as a growing cultural and moral disorientation. These concerns led to a movement to locate the now powerful medical sciences and technologies within the context of the humanities. These mid-20th century reflections had a continuity with what had in the late 19th and early 20th centuries engendered the Third[27] or New Humanism.[28] Abraham Flexner, for instance, in his 1928 Taylorian lecture argued that "the assessment of values, in so far as human beings are affected, constitutes the unique burden of humanism."[29] The popularity of C. P. Snow's Rede Lecture exploring the appropriate interplay of the culture of science and the culture of the humanities,[30] as well as the presence of movements to create a humanistic psychology and psychotherapy, reflected the growing felt need for cultural guidance.[31] However, attempts to institutionalize the medical humanities as the central source of this sought-for orientation for health care and the biomedical sciences failed.[32] It was bioethics that succeeded in garnering for itself a central social role and cultural moral normativity because bioethics was able to tie its intellectual concerns with the production of practitioners able to provide services for hospitals, Institutional Review Boards, and other organizations. Bioethics succeeded in combining the secular academic services of theologians with the ability to offer the secular equivalent of the pastoral care of chaplains and priests.

8.6 How Bioethics Succeeds, Even Though It Cannot Deliver What It Promised

The success of bioethics might seem to counter these doubts about bioethics due to its having failed to provide what it had promised: canonical moral guidance. People wanted morally normative direction and the wide acceptance of bioethics would seem to indicate that bioethics has succeeded in supplying useful moral advice. By the mid-1980s, bioethics programs were widely established in American medical schools, ethics committees were an ordinary part of hospital care, and courses in bioethics were offered on many college campuses.[33] An academic field had emerged replete with professors, encyclopedias, journals, and a growing body of scholarship.[34] However, closer examination indicates grounds for holding that, while bioethics promises one thing, it in fact succeeds in doing something else.

8 Why Clinical Bioethics So Rarely Gives Morally Normative Guidance   163

Bioethicists offer a complex bundle of services that constitute two distinct clusters of endeavors, one academic and the other practical or clinical. These two clusters of undertakings interplay in mutual support. As already noted, the clinical and public policy dimension of bioethics makes it plausible to create academic programs at universities and policy-making institutions. On the other hand, the academic base provides an intellectual *imprimatur* for the undertakings of clinical ethicists and like practitioners of bioethics. The place of bioethics in the academy supports the plausibility of clinical ethicists. The place of clinical ethicists in hospitals and elsewhere supports the plausibility for an academic recognition of the field.

It is generally taken for granted that academic bioethicists will disagree about substantive issues, yet the academic status and the authority of bioethics conveys a general status and authority to the field. As a part of humanities faculties, such persons are not expected to achieve agreement on substantive issues. On the other hand, clinical ethicists can affirm there is normative consensus in the field by providing non-normative guidance regarding the ethics that is established at law and in enforceable public policy. A confusion of the provision of the very effective non-morally-normative services usually offered by clinical bioethicists with the actual existence of a normative consensus may thus be created, suggesting that there must be more agreement at the theoretical and practical levels than in fact does exist. Clinical ethicists can thus establish themselves in hospitals by providing various forms of non-morally-normative services and counseling with both an academic imprimatur and the appearance of consensus.

As to the academic face of bioethics, bioethics provides cultural contributions similar to those of philosophers by bringing together analytic skills and a knowledge of the history of ideas. Academic bioethicists

(1) analyze ideas, concepts, and claims cardinal to an understanding of the moral issues raised by health care and the biomedical sciences,
(2) assess the soundness of arguments bearing on these issues, and
(3) provide geographies of different moral and moral-philosophical positions regarding such issues by showing how, given different initial premises and rules of evidence, one can establish different moral norms and viewpoints; thus, bioethicists show what act-utilitarians of a certain sort will affirm as appropriate choices, while showing that Kantians will embrace other choices.

Bioethics as such provides clarification and a better understanding of the intellectual moral context of health care and the biomedical sciences, but not actual canonical moral direction. Bioethicists disagree concerning the cardinal issues of their field, just as do moral philosophers generally. However, bioethics still has an intellectual plausibility if it can lay out the conceptual geographies of different moral or bioethical positions, along with different possible approaches to bioethical challenges. One need not expect any more coherence among the views of bioethicists than among those of moral philosophers.

Difficulties would become salient, were the clinical practitioners of secular bioethics generally to advance morally normative positions. Those who conceive of bioethics on the model of an established religious orthodoxy may expect moral advocacy, but they will be disappointed. Any concrete normative advice given must be advanced in the face of substantive disputes regarding the nature and foundations of morality and bioethics, as well as the content of canonical bioethical norms. From abortion, euthanasia, and health care reform, to cloning, the use of human embryonic stem cells, and the nature of disease (as for example regarding issues such as homosexuality), there are substantive disagreements. Clinical ethicists may clarify the nature of those disputes, but given a diversity of basic premises and rules of evidence in clinical bioethics, bioethicists cannot resolve substantive moral disputes. Despite insistent claims of consensus,[35] bioethics provokes major battles in the culture wars, fueled by clashing cultures and incompatible moral visions.[36] Thus, despite the aspirations of secular moral rationality to regard morality and bioethics *sub specie aeternitatis*, morality and bioethics fragment into numerous understandings of the morally reasonable. Again, given not just the persistence of moral pluralism but its salience, how can it be the case that bioethics has succeeded so well?

Secular bioethics succeeds because clinical bioethicists function in clinics and ethics committees, not primarily as normative ethicists, but instead through providing other, non-normative, moral services. Rather than doing what the term clinical ethicist might suggest bioethicists would generally do (i.e., give morally normative guidance) for the clinical conduct of health care, what they actually generally do is usually less controversial.[37]

Much of the advice they give concerns the ethos established at law and in public policy within a particular jurisdiction. The result is that with regard to clinical ethics consultations bioethicists operate in ways quite different from the way in which religious experts give advice within traditional religious communities structured by a thickly established orthodoxy. Secular clinical ethicists give guidance about an ethos established at law, including those who hire them successfully to navigate within the grey zones of that ethos, while also aiding in risk management, improved communications among patients, families, and physicians, the mediation of disputes, and risk management, while also providing clarification about the implications of particular moral commitments.

It turns out that the primary roles of clinical ethicists are closer to those of lawyers than to those of scientists or physicians, regarding whom one might have concern about how their employment by a particular agency, public or private, might corrupt their judgment. However, in the face of moral pluralism, clinical ethicists may properly be ethicists-for-hire, as attorneys are lawyers for hire, so as to plead a particular "ethics cause" within the grey zone of an ethos established at law. Within such a grey zone, "adversarial ethics" helps disclose the contours and boundaries of the ethos established at law. In the end, in the face of moral pluralism and within the constraints of the ethos established at law, or even in criticism of that ethos and in advocacy to establish a different ethos, legislatures, presidents, and pharmaceutical firms may properly hire those ethicists who will plead their views. One price of ineradicable secular moral pluralism is that one must foundationally rethink the role of clinical ethics.

8 Why Clinical Bioethics So Rarely Gives Morally Normative Guidance 165

The role of the clinical ethicist is thus rarely that of a moral adviser or moral guide. Clinical ethicists are not *phronemoi*. Clinical ethicists, health care ethics consultants, and bioethics consultants generally

(1) act as legal advisers who, since they are not lawyers admitted to the bar, can provide relatively inexpensive, on-the-spot legal advice about the ethos established at law and through public policy (e.g., on Institutional Review Boards they can provide information about relevant regulations and about how to apply that ethos);
(2) act as attorneys do in defending a particular interpretation of applicable law, in that there will always be grey zones of the ethos established at law and by public policy (as a result, ethicists for hire have no more of a conflict of interest than do attorneys when ethicists are hired to serve as ethics consultants for a corporation so as to defend a particular interpretation of grey zones in the legally established ethos; there is no conflict of interest in health care ethics consultants functioning as ethicists for hire (just as it would not be improper for lawyers) because their behavior is in accord with the nature of secular bioethics and secular ethics consultation);
(3) act as risk managers who help physicians and hospitals to lower the risk of malpractice actions through such practices as documenting the grounds for controversial clinical decisions and by their presence showing that there has been due diligence in considering what was at stake in problematic clinical cases;
(4) act as dispute mediators and communication specialists who aid parties such as physicians, patients, family members, and nurses better to collaborate in providing medical treatment, all of which is accomplished with the authority or imprimatur of being able to;
(5) act as ethics experts (i.e., anointed with the authority of bioethics as an academic moral-intellectual discipline), who possess special analytic knowledge and expertise about how to analyze what could be considered to be obligations in particular circumstances, given particular moral commitments and moral-theoretical assumptions (all of which is not normative advice), although actual normative moral guidance when given (e.g., as a Roman Catholic ethicist in a Roman Catholic hospital or as an advocate of a particular secular sect of morality and bioethics) is always in defense of one among many competing particular moral accounts or moral narratives.

This collage of services, this complex multi-tasking of bioethicists, has proven a useful contribution to hospitals and other health care facilities. The presence of clinical bioethicists demonstrates an institution's "caring concern" for moral issues and for the treatment of patients, while lowering moral and legal friction in the conduct of health care. The bioethicist functions as a useful cultural and moral factotum. The richness of this complexity and ambiguity challenges the American Society for Bioethics and Humanities to re-assess its attempt to articulate core competencies for healthcare ethics consultants (ASBH, 2011).

8.7 Rethinking Bioethics at the Beginning of the 21st Century

Against this background, one can appreciate the extraordinarily successful adaptation of the field of bioethics to the moral and cultural needs of the times. The astonishing success of bioethics shows how a service profession can function as a well-adapted social organism by combining numerous clinical services while protecting itself with an intellectual environmental camouflage that allows for a successful adaptation in different market environments. A bundle of different skills, roles, and facilities has been successfully brought together and given cultural standing by being nominally grounded in an academically accepted field. Despite the failure to achieve a canonical moral understanding, secular bioethicists have succeeded in binding together legal advice, liability risk management, and dispute mediation, while also offering the intellectual service of clarifying moral understandings, analyzing crucial concepts, assessing the coherence of arguments, and providing geographies of different possible moral and bioethical positions. Bioethics as this remarkable bundle of services has been exported across the world. In the face of moral diversity, although clinical ethicists cannot justifiably speak of a moral or ethical consensus, they act as if a moral or ethical consensus existed. In the face of deep and ongoing disputes, while being termed clinical ethicists, they serve as legal advisors, risk managers, and conflict mediators in ways such that their services provoke little controversy.

Along with the world-wide export of this model of bundling services, there has often also been an attempted export of the socio-democratic commitments and ideology, which characterized many (but surely not all) of the founders of American bioethics. Many of the founders of bioethics thought that their ideological commitments would be accepted as self-evident by others everywhere. These very particular cultural and ideological commitments are now provoking a reaction in the Pacific Rim against the American standard account of bioethics as a form of cultural colonialism (Fan, 2010, 2011). This has led to a critical reconsideration of the possibilities of bioethics.[38] Moral and bioethical pluralism, it should once again be stressed, is not just manifest internationally but is salient within individual societies and manifest in how different bioethical commissions and their recommendations are a function of who has political power.[39]

Bioethics has begun to recognize not just a plurality of moralities and moral visions, but the foundational circumstance that persons do not share the same basic moral premises, rules of inference, and understandings of human flourishing. There is a growing international appreciation of the consequences of the plurality of basic metaphysical commitments. The contrasts between positions are often stark. On the one hand, suffering, dying, and death are treated in secular bioethics as if everything came from no place, went to nowhere, and for no ultimate purpose,[40] while religious bioethics acknowledge ultimate points of reference. By default, much of secular bioethics gives central place to forbearance rights. In the face of conflicts among moral communities united around disparate content-full moralities and incompatible metaphysical narratives, there is little left for a global bioethics beyond a space for a procedural morality that can support the market, limited democratic polities,

8 Why Clinical Bioethics So Rarely Gives Morally Normative Guidance   167

and a rejection of content-full human claims rights.[41] The initial optimism of André Hellegers and Sargent Shriver along with their hopes for a canonical, morally normative bioethics has progressively been brought into question. However, bioethics as a hybrid of intellectual undertakings with a bundle of clinical services is likely to continue to flourish.

Notes

1. The term bioethics has, to say the least, a complex history. It appears to have been first coined by Fritz Jahr (1927) and then again independently crafted by Van Rensselaer Potter (1949), although his use in the 1970s (Potter, 1970, 1971) is often taken to be the first use of the term. Either André Hellegers (Reich, 1999) or Sargent Shriver (personal letter to H. Tristram Engelhardt, Jr., 26 January 2001) first recoined the term as it is now used.
2. The production of an encyclopedia of bioethics was taken to indicate the significance and status of the field (Reich, 1978); a second edition appeared in 1995, and a third edition in 2004 edited by Stephen Post. Since then, other overviews of the field have appeared. See, for example, Khushf (2004).
3. The literature exploring the genesis of bioethics is considerable. See Rothman (1991), Jonsen (1998), Stevens (2000), Abel (2001), and Walter and Klein (2003).
4. As recently as the late 20th century, many European countries maintained an established religion (e.g., England) or a number of established religions (e.g., Germany). In the second half of the 20th century, the standing of such established religions was in various ways and in different countries deflated, limited, or abolished.
5. The first Amendment to the compact styled the Constitution of the United States does not forbid the non-federal establishment of religion. It only forbids the establishment of a national, not a state, county, or city religion. "Congress shall make no law respecting an establishment of religion, or prohibiting the free exercise thereof; or abridging the freedom of speech, or of the press; or the right of the people peaceably to assemble, and to petition the government for a redress of grievances."
6. *Church of the Holy Trinity v. United States*, 143 US 457 (1892) at 470.
7. *United States v. Macintosh*, 283 US 605 (1931) at 625.
8. *Tessim Zorach v. Andrew G. Clauson et al.,* 343 US 306, 96 L ed 954, 72 S Ct 679 (1952) at 313.
9. "Evidence that Protestant Christianity [was] the functional common religion of [American] society would overwhelm us if we sought it out." Wilson (1986, p. 113). Huntington argues that the American ethos is rooted in Protestant Christianity (Huntington, 2004).
10. *Roy R. Torcaso v. Clayton K. Watkins*, 367 US 488, 6 L ed 2d 982, 81 S Ct 1680 (1961).
11. *School District of Abington Township v. Edward L. Schempp, William J. Murray et al. v. John N. Curlett et al.*, 374 US 203, 10 L ed 2d 844, 83 S Ct 1560 (1963).
12. Supreme Court rulings on abortion developed against the background of other holdings, particularly regarding access to contraception. *Griswold v. Connecticut*, 381 US 479, 85 S Ct 1678, 14 L Ed 2d 510 (1965); *Eisenstadt v. Baird*, 405 US 438, 92 S Ct 1029, 31 L Ed 2d 349 (1972); and *Roe v. Wade*, 410 US 113 (1973); and *In re Cruzan* 58 LW 4916 (June 25, 1990).
13. See http://www.atheists.org/
14. On April 22, 1864, the United States Congress authorized the motto "In God we trust" on American coinage. It appeared for the first time on the two-cent coin that year. The Coinage Act of February 12, 1873, states that "the Director of the Mint, with the approval of the Secretary of the Treasury, may cause the motto 'In God we trust' to be inscribed upon such coins as shall admit of such motto. . . ." Appendix to the Congressional Globe, February 12, 1873, Chap. CXXXI, Sec. 18, p. 237. As recently as July 30, 1956, the 84th Congress

168 H.T. Engelhardt, Jr.

declared that "The national motto of the United States is declared to be In God We Trust" (P.L. 84–140), Law 36 U.S.C. 186.

15. "The Treaties of Westphalia finally sealed the relinquishment by statesmen of a noble and ancient concept, a concept which had dominated the Middle Ages: that there existed among the baptized people of Europe a bond stronger than all their motives for wrangling—a spiritual bond, the concept of Christendom. Since the fourteenth century, and especially during the fifteenth, this concept has been steadily disintegrating.... The Thirty Years' War proved beyond a shadow of a doubt that the last states to defend the idea of a united Christian Europe were invoking that principle while in fact they aimed at maintaining or imposing their own supremacy. It was at Münster and Osnabrück that Christendom was buried. The tragedy was that nothing could replace it; and twentieth-century Europe is still bleeding in consequence" (Daniel-Rops, 1965, vol. 1, pp. 200–201).

16. For an account of the anti-religious violence of the French Revolution, see Vovelle (1991).

17. See Hofmann (1976), "Der Reichsdeputationshauptschluss," pp. 329–358; Eichendorff (1958). For an account of analogous cultural developments in Britain, see Wilson (1999).

18. Religious medical-moral-theological resources were considerable as one entered the mid-20th century, especially in Roman Catholicism. In the United States, Roman Catholic medical morality was in part articulated in ecclesial directives to American Roman Catholic hospitals. These were initially published as a code for the Catholic Health Association, which developed out of a movement (initiated in June, 1915) led by Charles B. Moulinier, S.J. Rev. Michael P. Bourke oversaw the creation of a code of ethics for the Diocese of Detroit in 1920, and in 1921 a surgical code was adopted by what was at that time styled the Catholic Hospital Association. Finally, a document with the title "Code of Ethics—1948" was written and first published in 1949. For an account of this history, see Griese (1987, pp. 1–19). At the same time, a literature developed exploring the nature of proper conduct under this code and for Roman Catholic physicians in general. See, for example, Kelly (1958). This work was initially published as pamphlets: Kelly (1949, 1950, 1951, 1953, 1954). See also Catholic Hospital Association (1971).

Until the mid-1960s, there was also a considerable and often quite sophisticated moral-theological literature available for Roman Catholic physicians, which grew out of an over 300-years Roman Catholic tradition of producing manuals to guide moral decisions. After the 2nd Vatican Council (1962–1965) this intellectual and scholarly tradition abruptly came to an end. In English, this literature included: Bonnar (1944), Bouscaren (1933), Capellmann (1882), Coppens (1897), Ficarra (1951), Finney (1922), Flood (1953–1954), Hayes et al. (1964), Healy (1956), Kenny (1952), La Rochelle and Fink (1944), McFadden (1946a, b), O'Donnell (1956), and Sanford (1905).

19. The Hippocratic corpus affirms a privileged status for the norms binding physicians, who are taken to be an established elite. "The Law", for instance, states that "Things however that are holy [meaning, as in the Hippocratic Oath, matters of the medical art] are revealed only to men who are holy [i.e., physicians]. The profane may not learn them until they have been initiated into the mysteries of science" (Hippocrates, 1959, vol. 2, p. 265).

20. The United States Supreme Court through a number of holdings recast the status of the medical profession from a quasi-guild into a trade. See, for example, *The United States of America, Appellants, v. The American Medical Association, A Corporation; The Medical Society of the District of Columbia, A Corporation; et al.*, 317 U.S. 519 (1943); and *American Medical Assoc. v. Federal Trade Comm'n*, 638 F.2d 443 (2d Cir. 1980).

21. At stake was a claim regarding clinical moral epistemology. Until the mid-20th century, many physicians acted with the conviction that they had come through their clinical experience to know what was morally at stake in the practice of medicine and how best to direct medical decisions. The view was that through practicing medicine a moral knowledge could be gained that was usually open only to physicians.

22. By the end of the 18th century, there was a considerable literature exploring medical morality. As this literature developed, it came ever more to constitute a medical morality articulated by

8 Why Clinical Bioethics So Rarely Gives Morally Normative Guidance   169

physicians for physicians. See, for example, Codronchi (1591), Castro (1614), Rau (1764), Frank (1779, 1976), Gregory (1770), and McCullough (1998a). For a critical edition of Gregory's work, see also McCullough (1998b). In the 19th century in the United States and Great Britain, this literature was augmented by the drafting of formal codes for professional medical deportment. See, for example, Percival (1803), Cartwright (1844), Medical Association (1839), and American Medical Association (1848). For some studies of these developments, see Konold (1962) and Baker (1995).

23. *Canterbury v. Spence*, 464 F. 2d 772, 789 (D.C. Cir. 1972); *Cobbs v. Grant*, 8 Cal. 3.d 229, 246; 502 P.2d 1, 12; 104 Cal. Rptr. 505, 516 (Calif. 1972); and *Sard v. Hardy*, 397 A. 2d 1014, 1020 (Md. 1977).

24. Rawls excludes a strong moral reading of his theory of justice; see Rawls' reflections on his account as applying only to a closed society, as well as his remarks on not taking a view *sub specie aeternitatis* (Rawls, 1971, pp. 8 and 587).

25. "[D]ie Eule der Minerva beginnt erst mit der einbrechenden Dämmerung ihren Flug" (Hegel, 1964, p. 37).

26. Traditional Christian theology, unlike secular philosophy, in that it is grounded in noetic experience, knows the meaning, direction, and significance of human cosmic history, namely, that it reaches from creation and the Fall through the Incarnation and Redemption to the Second Coming and the restoration of all things. Absent knowledge of the framing significance of cosmic and human history, humans are fundamentally disoriented, lost in the cosmos. Among other things, contemporary secular reflection severely limits answers to the third of Kant's famous three questions, "What may I hope for?" *Critique of Pure Reason* A805=B833.

27. The Third Humanism took many forms. See, for example, Rüdiger (1937) and Curtius (1932).

28. For a presentation of the scope and character of the New Humanism, see Hoeveler (1977) and Grattan (1968).

29. Flexner (1928, p. 12). Abraham Flexner (1866–1959) is best known for the Flexner Report concerning American medical education (Flexner, 1910).

30. Given concerns to find coherence of vision and moral orientation, Snow's explorations of the relationship between the culture of the humanities and the culture of the sciences attracted considerable attention. See Snow (1956). The original lecture printed as Snow (1962) saw a second edition with a substantial postscript, Snow (1964). For a 19th century Rede lecture anticipating Snow, see Arnold (1882). See, also, Lepenies (1985).

31. Many turned to psychology and psychotherapy to restore a humanistic perspective to a culture increasingly dominated by science and technology. See, for instance, Ellis (1973) and Schaefer (1978).

32. The aspiration to have humanism (usually invoked as an ambiguous notion) inform medicine led to the formation of the Society for Health and Human Values, which was later incorporated into the American Society for Bioethics and the Humanities. The Society for Health and Human Values, reflecting cultural forces that were salient prior to the emergence of bioethics, placed the humanities centrally, unlike the American Society for Bioethics and the Humanities, which came to give a more central focus on bioethics. "The Society was formally established in 1968 to encourage research, teaching and public interest in questions of human values as they arise in health and medical care. Its members are drawn from a wide spectrum of education, faculty members and health care practitioners who share a common interest in ethics, human values and the humanities in the health sciences" (Pellegrino and McElhinney, 1982, p. 1). See, also, Pellegrino (1979).

33. Interest in ethics committees gained momentum in part after an article cited in the Karen Quinlan decision (In the Matter of Karen Quinlan 70 NJ 10 [1976]) suggested that ethics committees were widespread (Teel, 1975).

34. A wide range of encyclopedias, handbooks, and summaries of bioethics now exists: Leone and Privitera (1994), Hottois and Missa (2001), Lecaldano (2002), Korff et al. (2002), Copray (2004) and Russo (2004).

35. There is no consensus about what constitutes a moral or bioethical consensus, because there is no adequate morally normative answer to the question as to what amount of agreement about what moral issues by whom should be morally compelling and why. Behind this question is the challenge of determining who should count as moral and bioethical experts. See Bayertz (1994), ten Have and Sass (1998), and *Cambridge Quarterly* (2002).
36. The profound conflicts between disparate cultural and moral perspectives have been the focus of numerous studies. See Hunter (1991) and Huntington (1998).
37. On the issue of the multi-form character of clinical ethics, I am indebted to years of conversations with Lisa Rasmussen. See Rasmussen (2003). For a more complete presentation of my views on these issues, see Engelhardt (2003b).
38. The hegemony of American bioethics is now challenged by an attempt to think through bioethics anew. A critical assessment of the bioethics established in the 1970s and 1980s is emerging, especially in Asia. In particular, there is a move to establish a bioethics that does not share the presuppositions of the standard American account. See, for example, Alora and Lumitao (2001), Fan (2010, 2011), Hoshino (1997), and Qiu (2004).
39. The ideologically partisan character of bioethics can be appreciated by comparing the differences between the reports issued by Bill Clinton's National Bioethics Advisory Commission versus those from George W. Bush's President's Council on Bioethics. This is particularly clear with respect to statements regarding cloning and reproduction. See, for example, National Bioethics Advisory Commission (1997, 1999, 2000). Contrast that with the President's Council on Bioethics (2002, 2003, 2004a, b).
40. Without a canonical point of ultimate reference, moral and metaphysical perspectives lose a uniting regulative point of focus and fragment. "Atheism appears in this light as another catastrophic Tower of Babel" (Vattimo, 1991, p. 31).
41. In the face of moral diversity, morality de facto can take on a two-tier character, such that one tier recognizes that there is moral truth, while the other tier supports procedures for collaboration among moral strangers in the face of irreducible moral pluralism and disagreement. This second tier does this by stepping back from attention to content-full moral truth and relying instead on collaboration based on mere agreement or permission. Given the limits of discursive moral rationality and the deafness of many to the claims of God, the second tier is all that is available for many when they seek to collaborate peaceably. (See Engelhardt, 1996.)

References

Abel i Fabre, F., S.J. 2001. *Bioética: Origenes, Presente y Futuro*. Madrid: Mapfre.
Alora, A. tan Alora, and J.M. Lumitao. eds. 2001. *Beyond a Western bioethics: Voices from the developing world*. Washington, DC: Georgetown University Press.
American Medical Association. 1848. *Code of medical ethics adopted by the American Medical Association at Philadelphia in May, 1847, and by the New York Academy of Medicine in October, 1847*. New York: H. Ludwig.
American Society for Bioethics and Humanities. 2011. *Core competencies for healthcare ethics consultation, 2nd ed.* Glenview, IL: American Society for Bioethics and Humanities.
Arnold, M. 1882. *Literature and science*. Cambridge: Cambridge University Press.
Baker, R. ed. 1995. *The codification of medical morality: Historical and philosophical studies of the formalization of Western morality in the 18th and 19th centuries*, vol. 2: *Anglo-American medical ethics and medical jurisprudence in the 19th century*. Dordrecht: Kluwer.
Bayertz, K. ed. 1994. *The concept of moral consensus: The case of technological interventions into human reproduction*. Dordrecht: Kluwer.
Bonnar, A. 1944. *The Catholic doctor*. London: Burns Oates & Washbourne.
Bouscaren, T.L. 1933. *Ethics of ectopic operations*. Chicago: Loyola University Press.
Cambridge Quarterly of Healthcare Ethics 11 (Winter 2002): 1–108.

8 Why Clinical Bioethics So Rarely Gives Morally Normative Guidance

Capellmann, C.F.N. 1882. *Pastoral medicine* (trans: Dassel, William). New York: F. Pustet; orig. 1877.

Cartwright, S.A. 1844. Synopsis of medical etiquette. *New Orleans Medical and Surgical Journal* 1.2: 101–104.

Castro, R. 1614. *Medicus-Politicus: Sive de officiis Medicopoliticis Tractatus*. Hamburg: Frobeniano.

Catholic Hospital Association. 1971. *Ethical and religious directives for Catholic health facilities*. St. Louis, MO: Catholic Hospital Association.

Codronchi, G. 1591. *De Christiana ac tuta medendi ratione libri duo*. Ferrara: Benedictum Mammarellum.

Coppens, C. 1897. *Moral principles and medical practice*, 3rd ed. New York: Benziger Brothers.

Copray, N. ed. 2004. *EthikJahrbuch 2004. Von Bioethik bis Weltethos, von Freiheitsethik bis Unternehmensethik*. Frankfurt/M: Fairness-Stiftung.

Curtius, E.R. 1932. *Deutscher Geist in Gefahr*. Stuffgart: Deutsche Verlags-Anstalt.

Daniel-Rops, H. 1965. *The Church in the seventeenth century (Le grand siècle des âmes* [1963]), 2 vols. (trans: Buckingham, J.J.). Garden City, NY: Doubleday.

Dupré, L. 1993. *Passage to modernity*. New Haven, CT: Yale University Press.

Eichendorff, J.F. von. 1958. 'Über die Folgen von der Aufhebung der Landeshoheit der Bischöfe und der Klöster in Deutschland.' In *Werke und Schriften*, vol. 4, 1133–1184. Stuttgart: Cotta'sche.

Ellis, A. 1973. *Humanistic psychotherapy*. New York: McGraw Hill.

Engelhardt, H.T., Jr. 1996. *The foundations of bioethics*, 2nd ed. New York: Oxford University Press.

Engelhardt, H.T., Jr. 2000. *The foundations of christian bioethics*. Salem, MA: Scrivener Publishing.

Engelhardt, H.T., Jr. 2002. 'The ordination of bioethicists as secular moral experts.' *Social Philosophy & Policy* 19(Summer): 59–82.

Engelhardt, H.T., Jr. 2003b. The bioethics consultant: Giving moral advice in the midst of moral controversy. *Healthcare Ethics Committee Forum* 15(December): 362–382.

Fan, R. 2010. *Reconstructionist Confucianism*. Dordrecht: Springer.

Fan, R. ed. 2011. *The renaissance of Confucianism in contemporary China*. Dordrecht: Springer.

Ficarra, B.J. 1951. *Newer ethical problems in medicine and surgery*. Westminster, MD: Newman Press.

Finney, P.A. 1922. *Moral problems in hospital practice*, 2nd ed. St. Louis, MO: B. Herder.

Flexner, A. 1910. *Medical education in the United States and Canada, a report to the Carnegie Foundation for the Advancement of Teaching*, Bulletin No. 4. New York: Carnegie Foundation.

Flexner, A. 1928. *The burden of humanism*. Oxford: Clarendon Press.

Flood, P. 1953–1954. *New problems in medical ethics*, 2 vols (trans: Carroll, Malachy Gerald). Westminster, MD: Newman Press.

Frank, J.P. 1779. *System einer vollständigen medicinischen Polizey*. Mannheim: C.F. Schwan.

Frank, J.P. 1976. *A system of complete medical policy* (trans: Wilim, E.). Baltimore, MD: Johns Hopkins University Press.

Gay, P. 1995. *The enlightenment: The rise of modern paganism*. New York: W.W. Norton.

Grattan, C.H. ed. 1968. *The critique of humanism*. Freeport, NY: Books for Libraries Press.

Gregory, J. 1770. *Observations on the duties and offices of a physician*. London: Strahan.

Griese, O.N. 1987. *Catholic identity in health care: Principles and practice*. Braintree, MA: Pope John Center.

Hayes, E., P. Hayes, and D. Kelly. 1964. *Moral principles of nursing*. New York: Macmillan.

Healy, E.F. 1956. *Medical ethics*. Chicago: Loyola University Press.

Hegel, G.W.F. 1964. *Grundlinien der Philosophie des Rechts*. Stuttgart: Friedrich Frommann Verlag.

Hegel, G.W.F. 1970. *Hegel's philosophy of nature* (ed. and trans: Petry, M.J.). London: George Allen & Unwin.

Hippocrates. 1959. 'The law.' In *Hippocrates*, vol. 2 (trans: Jones, W.H.S.), 262–265. Cambridge, MA: Harvard University Press.

Hoeveler, J.D., Jr. 1977. *The new humanism*. Charlottesville: University Press of Virginia.

Hofmann, H.H. ed. 1976. *Quellen zum Verfassungsorganismus des heiligen römischen Reiches deutscher Nation*. Darmstadt: Wissenschaftliche Buchgesellschaft.

Hoshino, K. ed. 1997. *Japanese and Western bioethics*. Dordrecht: Kluwer.

Hottois, G., and J.-N. Missa. eds. 2001. *Nouvelle Encyclopédie de Bioéthique*. Brussels: DeBoeck & Larcier.

Hume, D. 1932. *The letters of David Hume*, 2 vols. Oxford: Clarendon Press.

Hunter, J.D. 1991. *Culture wars: The struggle to define America*. New York: Basic Books.

Huntington, S.P. 1998. *The clash of civilizations and the remaking of world order*. New York: Simon & Schuster.

Huntington, S.P. 2004. *Who are we: The challenges to America's national identity*. New York: Simon & Schuster.

Jahr, F. 1927. Bio-Ethik. Eine Umschau über die ethischen Beziehungen des Menschen zu Tier und Pflanze. *Kosmos. Handweiser für Naturfreunde* 24: 2–4.

Jonsen, A.R. 1998. *The birth of bioethics*. New York: Oxford University Press.

Kelly, G.S.J. 1949. *Medico-moral problems, Part I*. St. Louis, MO: Catholic Hospital Association.

Kelly, G.S.J. 1950. *Medico-moral problems, Part II*. St. Louis, MO: Catholic Hospital Association.

Kelly, G.S.J. 1951. *Medico-moral problems, Part III*. St. Louis, MO: Catholic Hospital Association.

Kelly, G.S.J. 1953. *Medico-moral problems, Part IV*. St. Louis, MO: Catholic Hospital Association.

Kelly, G.S.J. 1954. *Medico-moral problems, Part V*. St. Louis, MO: Catholic Hospital Association.

Kelly, G.S.J. 1958. *Medico-moral problems*. St. Louis, MO: Catholic Hospital Association.

Kenny, J.P. 1952. *Principles of medical ethics*. Westminster, MD: Newman Press.

Khushf, G. ed. 2004. *Handbook of bioethics*. Dordrecht: Kluwer.

Konold, D.E. 1962. *A history of American medical ethics, 1847–1912*. Madison, WI: State Historical Society of Wisconsin.

Korff, W., L. Beck, and P. Mikat. eds. 2002. *Lexikon der Bioethik*. Gütersloh: Gütersloher Verlagshaus.

Krause, E.A. 1996. *Death of the guilds*. New Haven, CT: Yale University Press.

La Rochelle, S.A., and C.T. Fink. 1944. *Handbook of medical ethics* (trans: Poupore, M.E.). Westminster, MD: Newman Book Shop.

Lecaldano, E. 2002. *Dizionario di bioetica*. Rome: Laterza.

Leone, S., and S. Privitera. eds. 1994. *Dizionario di Bioetica*. Bologna: EDB-ISB.

Lepenies, W. 1985. *Die drei Kulturen*. Munich: Hanser.

Lyotard, J.-F. 1984. *The postmodern condition* (trans: Bennington, G., and B. Massumi). Manchester: Manchester University Press.

McCullough, L.B. 1998a. *John Gregory and the intention of professional medical ethics and the profession of medicine*. Dordrecht: Kluwer.

McCullough, L.B. ed. 1998b. *John Gregory's writings on medical ethics and philosophy of medicine*. Dordrecht: Kluwer.

McFadden, C.J. 1946a. *Medical ethics for nurses*. Philadelphia, PA: Davis.

McFadden, C.J. 1946b. *Medical Ethics*. Philadelphia, PA: Davis.

Medical Association of North Eastern Kentucky. 1839. *A system of medical etiquette*. Maysville, KY: Maysville Eagle.

National Bioethics Advisory Commission. 1997. *Cloning human beings*, vol. 1, vol. 2, and Executive Summary. Rockville, MD: National Bioethics Advisory Commission.

National Bioethics Advisory Commission. 1999. *Ethical issues in human stem cell research*, vol. 1, and Executive Summary. Rockville, MD: National Bioethics Advisory Commission.

National Bioethics Advisory Commission. 2000. *Ethical issues in human stem cell research*, vol. 2. Rockville, MD: National Bioethics Advisory Commission.

O'Donnell, T.J. 1956. *Morals in medicine*. Westminster, MD: Newman Press.

Oleksa, Michael. 1992. *Orthodox Alaska,* 171–186. Crestwood, NY: St. Vladimir's Seminary Press.

Pellegrino, E.D. 1979. *Humanism and the physician.* Knoxville, TN: University of Tennessee Press.

Pellegrino, E.D., and T. McElhinney. 1982. *Teaching ethics, the humanities, and human values in medical schools.* Washington, DC: Society for Health and Human Values.

Percival, T. 1803. *Medical ethics.* Manchester: Russell.

Potter, V.R. 1949. *Global bioethics – building on Leopold's legacy.* East Lansing, MI: Michigan State University Press.

Potter, V.R. 1970. Bioethics: The science of survival. *Perspectives in Biology and Medicine* 14: 127–173.

Potter, V.R. 1971. *Bioethics: Bridge to the future.* Englewood Cliffs, NJ: Prentice Hall.

President's Council on Bioethics. 2002. *Human cloning and human dignity.* New York: PublicAffairs.

President's Council on Bioethics. 2003. *Beyond therapy.* New York: Dana Press.

President's Council on Bioethics. 2004a. *Monitoring stem cell research.* Washington, DC: President's Council on Bioethics.

President's Council on Bioethics. 2004b. *Reproduction and responsibility.* Washington, DC: President's Council on Bioethics.

Qiu, Ren-Zong. ed. 2004. *Bioethics: Asian Pperspectives, a quest for moral diversity.* Dordrecht: Kluwer.

Rasmussen, L. 2003. *Clinical bioethics: Analysis of a practice.* Unpublished doctoral dissertation, Rice University, Houston, TX.

Rau, W.T. 1764. *Gedanken von dem Nutzen und der Nothwendigkeit einer medicinischen Policeyordnung in einem Staat.* Ulm: Stettin.

Rawls, J. 1971. *A theory of justice.* Cambridge, MA: Harvard University Press.

Rawls, J. 1985. Justice as fairness: Political not metaphysical. *Philosophy and Public Affairs* 14: 223–251.

Rawls, J. 1993. *Political liberalism.* New York: Columbia University Press.

Reich, W.T. ed. 1978. *The encyclopedia of bioethics.* New York: Macmillan Free Press.

Reich, W.T. 1999. 'The 'wider' view: André Hellegers' passionate, integrating intellect and the creation of bioethics.' *Kennedy Institute of Ethics Journal* 9: 25–51.

Rothman, D.J. 1991. *Strangers at the Bedside.* New York: Basic Books.

Rüdiger, H. 1937. *Wesen und Wandlung des Humanismus.* Hamburg: Hoffman & Campe.

Russo, G. ed. 2004. *Enciclopedia di Bioetica e Sessuologia.* Torino: Editrice Elledici.

Sanford, A. 1905. *Pastoral medicine: A handbook for the catholic clergy.* New York: Joseph Wagner.'

Schaefer, J.B.P. 1978. *Humanistic psychology.* Englewood Cliffs, NJ: Prentice Hall.

Schneewind, J.B. 1998. *The invention of autonomy.* New York: Cambridge University Press.

Snow, C.P. 1956. The two cultures. *New Statesman* 6 Oct.: 413–414.

Snow, C.P. 1962. *The two cultures and the scientific revolution.* Cambridge: Cambridge University Press.

Snow, C.P. 1964. *The two cultures: And a second look.* Cambridge: Cambridge University Press.

Stevens, M.L.T. 2000. *Bioethics in America.* Baltimore, MD: Johns Hopkins University Press.

Teel, K. 1975. The physician's dilemma: a doctor's view: What the law should be. *Baylor Law Review* 27: 6–9.

ten Have, H., and H.-M. Sass. eds. 1998. *Consensus formation in healthcare ethics.* Dordrecht: Kluwer.

Vattimo, G. 1991. *The end of modernity.* Baltimore, MD: Johns Hopkins University Press.

Vovelle, M. 1991. *The revolution against the church* (trans: José, Alan). Columbus, OH: Ohio State University Press.

Walter, J.K., and E.P. Klein. eds. 2003. *The story of bioethics*. Washington, DC: Georgetown University Press.

Wilson, J. 1986. 'Common religion in American society.' In *Civil religion and political theology*, ed. L.S. Rouner, 111–124. Notre Dame, IN: University of Notre Dame Press.

Wilson, A.N. 1999. *God's funeral*. New York: W.W. Norton.

Part III
The Incredible Search for Bioethical Professionalism: Some Final Critical Reflections on Circular Thinking

Chapter 9
On the Social Construction of Health Care Ethics Consultation

Jeffrey P. Bishop

The story that medicine tells about itself is a story of medical progress. The story goes something like this; we once lived in darkness with diseased bodies until great men of science opened our eyes and shed light into the cave of human ignorance. In this story, every generation in medicine makes strides forward pushing back the boundaries of ignorance. Sometimes the strides are great; sometimes the strides are small. Always the movement is forward toward some version of utopia, whether that be progress toward an idealized human immortality, as suggested by Daniel Callahan (medical science), or progress in stamping out all human suffering and promoting individual human well-being (medical practice). Though this history is never written down, it nonetheless animates medicine's self-image.

The same seems to be true with bioethics generally. When evil men of science began to deploy their technologies on patients that didn't want them, bioethics stepped in and righted the relationship that had gone askew. When medicine began to bump up against serious moral questions presented by its technologies, bioethics helped to steer a path that preserved some semblance of a morally sensitive medicine. Medicine had always had an ethics not dissimilar to the surrounding culture within which Western medicine arose (McCullough, 1998; Jonsen, 1998), yet two features of modern life led to challenges of these robust and particular ethics; the first was the rise of technology (Callahan, 2000) and the second was the secularization of the West. (Engelhardt, 1996, 1991) The patient's bedside became the battleground of various culture wars, requiring someone to arbitrate. It seemed that technological development had so far outrun the moral development, whatever that might be, that we needed first theologians and then philosophers and, today, clinical ethicists. These thinkers and practitioners began to span the gap between the technological and the moral alleviating the moral arrogance of modern medicine; or so the story goes. Bioethics fancies itself as the field that offers reasoned and reasonable justification for medical advancements in science; clinical ethicists offer

---

J.P. Bishop (✉)
Albert Gnaegi Center for Health Care Ethics, Saint Louis University, Salus Center, Room 527,
3545 Lafayette Ave, St. Louis, MO 63104, USA
e-mail: jbisho12@slu.edu

H.T. Engelhardt, Jr. (ed.), *Bioethics Critically Reconsidered*,
Philosophy and Medicine 100, DOI 10.1007/978-94-007-2244-6_9,
© Springer Science+Business Media B.V. 2012

reasoned and reasonable reflection on the deployment of those technologies at the patient's bedside. The latter now hope to ensconce in policy and procedure, as well as in new institutions, a set of skills to facilitate between the patient and the health care team, a skill-set that all clinical ethicists should deploy in order to keep medical practitioners on the straight and narrow.

Elsewhere (Bishop et al., 2009) I, along with my colleagues, have shown how this story of progress is wrapped into the bureaucratic processes of medicine. By appeal to an evidence base—a value held in high regard by medical practitioners—and by appeal to the dictates of the Clinical Ethics in particular, as a subset of bioethics, thinks of itself as the conscience of the institution (Frolic and Chidwick, 2010, p. 20) trying to get institutions to do the right thing, or as the midwife producing good and humane outcomes through its processes of engagement and facilitation. All the while, the field also wants to suggest that it does not offer moral expertise.

At the October 2009 meeting of the American Society for Bioethics and Humanities, a panel discussion "Clinical Ethics Consultant: Should it Be a Certified Profession?," someone stood up at the end of the meeting outraged that Clinical Ethicists—or as I shall call them here, Health Care Ethics (HCE) consultants— were not already credentialed and certified. In fact, the questioner was a clinical ethics consultant (and physician) and he stated something to the effect of, "We have been debating this question for years and it is time to act. There are unethical things going on in small community hospitals all over this country and we need certified clinical ethics consultants to help stop these behaviors." The speaker's statement struck me as funny on many different levels, given that the mantra for clinical ethics consultants is that they are not moral or ethical experts. If clinical ethicists are not moral experts, how could our questioner possibly know that something unethical was going on in small community hospitals? If clinical ethics is not about ethical or moral expertise, then in what way could a clinical ethicist possibly stop the "unethical behaviors" purportedly rampant throughout small community hospitals?

9.1 Goals or Goods

It is difficult to say what Health Care Ethics Consultation (HCEC) is. It is clear that it is some sort of practice, though precisely what sort is not clear. Unlike the traditional practices of nursing, medicine, law, and pastoral care, there is no clear good at which HCEC aims. The practices of nursing and medicine are aimed at health and human flourishing, for the most part. The practice of law is aimed at continued freedom and prospering of particular citizens in the face of various threats from both the state and other citizens. The practice of pastoral care, while once aimed at the salvation of souls or at supporting the supplicant's faithfulness in the face of trial, is still aimed at some sort of spiritual good, even if generic chaplains have difficulty articulating exactly what spirituality is after it has been separated from its robust metaphysical moral roots. Fortunately for the faithful, the parish priests, pastors, and rabbis have not been cut off from these roots for the most part. The point is that the practices of nursing, medicine, law, and pastoral care at least historically have been tied to rich

9 On the Social Construction of Health Care Ethics Consultation

metaphysical moral traditions, and even while those in health care have repeatedly thinned out the metaphysical moral language, the goods attendant to their practices are still articulated with some degree of metaphysical and moral import.

Yet, HCEC appears only to have a set of goals, but no a clear set of goods that it promotes. The goals articulated by clinical ethics consultants are in some way related to the goods of medical care. So, I will attempt to articulate how medicine negotiates its goods, and where HCEC might relate to the goods of medical practice. Medicine, as a good of society, has spent hundreds of years exploring and articulating their goods, where clinical ethics has only bee around for three or four decades.

The clinical services housed in our healthcare institutions have interdependent relationships that require ongoing negotiations of the goods that define their shared interests. For example, a single doctor might spend three hours forming a strong relationship with a particular patient. Yet no medical practice can financially sustain a doctor's spending this much time with a patient. It is true that after spending so much time with a single patient, the physician may have a nuanced understanding of the patient's moral values and a thorough understanding of her patient's medical condition. Yet the doctor will go broke and lose money for his clinic or hospital. Thus, healthcare institutions must negotiate with doctors for how time will be distributed among patients in relation to other relevant goods, such as strong relationships with patients. The good of strong relationships is a good internal to the practice of medicine, but goods external to the practice, such as financial stability for the doctor and for the healthcare institution are always balanced against the goods internal to the practice of medicine.

Our institutions trade in external goods, that is to say, goods not central to the practices themselves. As noted by Alasdair MacIntyre, "They [social institutions] are involved in acquiring money and other material goods; they are structured in terms of power and status, and they distribute money, power and status as rewards. Nor could they do otherwise if they are to sustain not only themselves but also practices of which they are a bearer." (MacIntyre, 1984, p. 194). Put differently, while institutions create, sustain, and perpetuate practices, they also pose threats to the goods internal to practices, such as the good of spending time with a patient. However, physicians will be penalized if they adhere too stringently to the good of knowing patients extremely well. And moreover, they will be awarded for efficiency, a good external to the practice, but also a necessary condition for practice. Institutions control the distribution of external goods such as money, status, and power, and thus service providers have incentive to conform to institutional standards, sometimes at the expense of goods internal to the practice. In short, the goods of institutions and the goods of practices are intertwined, but they must be identified as both supportive and threatening to one another.

Clinical ethics is parasitic upon medical practice and dependent on the set of goods internal to health care. HCEC can be understood as a good of medicine, but only insofar as HCEC assists medicine in carrying out its goods—medicine's goods themselves being in complex relations with other institutions and practices. Yet HCEC's relationship to medicine is precarious, because at the same time, HCEC brings to bear other goods, taken from other institutions and concepts that are central to the identities of those who claim the mantle of HCE Consultant.

Let's drill down a little deeper. It is typically good for doctors to use therapies for which there is some empirical evidence, or at least in the absence of evidence, some theoretical justification for an intervention. The support is even stronger if the prevailing medical opinion is that a particular treatment will have some level of reasonable effectiveness, at some level of acceptable risk and cost. So, doctors will make judgments based on evidence, theories, professional standards of care, and after balancing the risks and benefits of both doing and not doing the intervention.

Yet, the goods of theoretical support, evidentiary support, and prevailing medical opinion are all goods directed at another good, the patient's health. Thus, under a medical paradigm, justice requires what is proportionate to the various goods of theoretical or empirical justification aimed at some reasonable understanding of what human thriving is or ought to be, and the chances of achieving some degree of flourishing. So, what happens when the physicians determine that a standard intervention does not exist; and what if the evidence base is poor; and what if there is little theoretical support for the intervention; and what if the chances of success are extremely thin or the intervention very risky? If the intervention fails in one or in any combination of these areas of practical judgment, the physician could easily decide not to intervene.

Yet, there might be a patient who demands a procedure or intervention that the physicians deem unnecessary, or theoretically not justified, or too risky relative to the hope for benefit. The goods as understood by the patient, then, might come into conflict with the goods internal to the practice of medicine. The patient might claim that the doctors are violating her autonomy; or she might claim that she has a human right to a particular intervention, thus challenging the goods internal to medical practice. She can legitimately claim that it is her life, and the doctors cannot possibly understand her best interests and the goods of her own life that are utterly particular to her. Still, her own self-understanding of her good is in relation to other goods external to her life, such as the goods from larger social institutions and legal traditions—some ancient, some more recent. So, when the goods of her life are perceived to be in conflict with the goods as articulated by medicine, there are resources that she has at her disposal to challenge the doctor's claim that a therapy is not indicated, and is not "good" for her. She can appeal to another doctor, or to a law-suit, or to the media, if the goods thought to be central to her life are thought to be at stake. In the hospital setting, she can also appeal to the HCE consultant.

It is here in a kind of secondary position that HCEC comes into play. But, what exactly are the goods internal to the practice of clinical ethics consultation? How do they relate to the goods of medicine? How do the goods of HCE consultants relate to patient goods, or the goods of the consultant's employer? Are the goods promoted by HCE consultants really goods best supported in the legal system, which has always been the social mechanism to arbitrate competing goods? In the process-driven guidelines of HCEC, the consultant is supposed to clarify the goals of care. But are goals the same as goods? Without a normative set of goods, HCEC can only articulate the goals as derivative of the goods articulated by the patient, or by medicine, or by the law, or by their employers. Thus, HCEC is directed at means to achieve goals articulated by the patient or by the professions of law or medicine. HCEC is thus parasitic upon other practices with more robust goods.

9.2 The Goals of HCEC

Between February 2006 and November 2009, a task force of the American Society for Bioethics and Humanities (ASBH) worked to revise and update The Core Competencies for Health Care Ethics Consultation. After a few months of commentary, a final version was to be published. However, the official second edition has not yet been made available as of February 2011. Thus, my commentary refers to the draft version of the Core Competencies.

This document refers to goals, but it does not clarify what the goods of HCEC are. What goods do the HCE consultant aim at such that HCE consultants can claim to be a bona fide practice? The Core Competencies draft document states:

> While all health care providers engage in ethical decision-making as part of their everyday practice (e.g. in facilitating informed consent with a patient or family before a procedure), health care ethics (HCE) consultants (or "ethics consultants") differ from other health care providers in that they have been assigned by their institutions the distinctive role of responding to specific ethical concerns and questions that arise in the delivery of health care, and therefore require a distinctive set of competencies to effectively perform this role. (Core Competencies, 4)

Several important points come into relief here. Whatever the goods of HCEC might be, they are derivative of the institutions of medical practice, or the hospital institution that supports them. At least, this is what the Core Competencies seem to be claiming. Yet, HCE consultants don't think of themselves as upholding the goods of medicine or of the goods of their employers. (Frolic) So, they must be upholding the values and goods of the patient. Are they not then best thought of as patient advocates? And if they are patient advocates in the face of the overwhelming powers of the health care and medical institutions (with their separate goods), would the goods of health for a particular patient not be best served by lawyers? And is there no conflict of interest, given that the vast majority of HCE consultants are employed by hospitals?

Still HCE consultants want to see themselves as different from lawyers. They still want to uphold the distinction between law (political philosophy) and ethics (moral philosophy). "Ethics consultation is an expert response to one or more specific ethics questions posed to a HCE consultant—that is, questions about which decisions are right or which actions should be taken when there is uncertainty or conflict about values." (Core Competencies, 5) Yet, of course, HCE consultants do not want to think of themselves as moral experts, but as experts of ethical (as opposed to legal) analysis. (Core Competencies, 6) The general "goal is more likely to be achieved if consultation accomplishes the intermediary goals." (Core Competencies, 6) These intermediary goals are to "indentify and analyze" value uncertainty, and to "facilitate resolution" with special attention "to the interests, rights, and responsibilities of all those involved." (Core Competencies, 6) The drafters of the Core Competencies document continue:

> Successful HCEC will also serve the goals of helping to promote practices consistent with ethical norms and standards; informing institutional efforts at policy development, quality improvement, and appropriate utilization of resources by identifying the causes of ethical concerns; and assisting individuals and the institution in handling current and future ethical problems by providing education in health care ethics. (Core Competencies, 6)

What seems to be clear is that HCEC promotes goals aimed at achieving medical goods, as well as patient goods, but it is unclear what the goods of HCEC are per se. It is also unclear what counts as "ethical concerns." Goals are process oriented; goods are a bit more illusive Yet HCEC seems to be directed at the processes rather than at the goods. In other words, HCEC is aimed at means, rather than at goods. Thus, the Core Competencies focus mostly on processes and "intermediary goals," or rather instrumental goals, and less on the goods usually attendant to robust practices, such as medicine or law.

9.3 Establishing the Right Model

Thus, HCEC aims at skills and processes rather than goods or ends. As with any group that has to create itself de novo, HCEC has to establish itself against a bogyman. As noted by Michel Foucault, one's status is always greater when one has to establish oneself in the face of an enemy. (Foucault, 1994, p. 125) In this instance, there are two enemies, the authoritarian clinical ethicist and the pure consensus clinical ethicist. (Core Competencies, 9) The best approach falls "between one extreme that might be termed the "authoritarian approach" and another that might be termed the "pure consensus" approach." (Core Competencies, 9) The best approach is of course, the "ethics facilitation" approach. Thus, the draft document of the Core Competencies states that there are three approaches to clinical ethics; or rather, there are multiple approaches between these extremes, with the correct approach—the ethics facilitation approach—sliding between these two extreme poles. Acknowledging the false dichotomy in a footnote, the draft document describes the correct model for clinical ethics consultation. (Core Competencies, 10) So, the standard and endorsed approach is only possibly illustrated between two falsehoods.

The Core Competencies also attempt to define ethics consultation by defining its boundaries—what it is not. Those who request consultation, "may sometimes contact the ethics consultation service seeking assistance primarily with concerns that are better handled by other mechanisms within the organization that are established to handle such things as general complaints, allegations of misconduct, or requests for medical opinions, legal advice, or spiritual support." (Core Competencies, 8) So, perhaps we can say what kind of practice HCEC is by virtue of showing what it is not. It is not patient advocacy (at least not as "patient advocates" are understood by hospital institutions). HCEC does not provide legal or spiritual advice, or medical

9 On the Social Construction of Health Care Ethics Consultation 183

or professional advice. The document does not say exactly what kind of advice the HCE consultants give.

In addition, the framers of the document also want us to believe that there are generally accepted standards upon which HCEC ought to be founded. Statements like this are peppered throughout the document: "Some generally agreed-upon standards for HCEC services are listed here." (Core Competencies, 14) As proof of the generally accepted standards, the Core Competencies document cites a document created and published by the National Center for Ethics in Health Care located in the Veterans Health Administration (VHA). (Fox et al., 2006) The publication that is cited is actually a document dictating to local VHA hospitals and their ethics consultation services the standards for clinical ethics consultation. (Fox et al., 2006, p. ii) It also seems to have pretensions in dictating the standards for ethics consultation for all HCE consultants; or at least the drafters of the document seem to think that these standards for Federal Employees of the VHA ought to be utilized by all consultants. One would expect that a citation to the "generally agreed-upon standards" for clinical ethics consultation would be a national survey of those who participate in clinical ethics consultation, or a consensus statement by an austere group of clinical ethicists. Oddly enough, the sort of document to which one would point as documentation of the "generally agree-upon standards" for ethics consultation would be something akin to the Core Competencies document itself as opposed to a statement of the generally agreed-upon standards of the National Center for Ethics in Health Care, a VHA group.

The document repeatedly mentions ethical standards, but neither these standards, nor their ethical or philosophical grounding is articulated. Instead, we find procedural standards. There is no ethical content; they emphasize the "emerging procedural standards;" these standards are:

1. Open access
2. Sound ethics consultation process
3. Notification of a case consult
4. Adequate documentation
5. Comprehensive policy
6. Evaluation, quality review and improvement (Core Competencies, 14)

Open access refers to the idea that anyone or almost anyone can place a consult. The largest section on these standards is devoted to Sound Ethics Consultation Process, which focuses on the importance of having a process. In rather circular fashion, the Sound Ethics Consultation Process should have clearly delineated processes by which one goes about the consultation process, including how to hold meetings. The key stakeholders, including patient, patient's family, as well as key parties to patient care, should be notified that a consult has been called. In addition, there should be adequate documentation of the process followed, and the decisions made during the consultation. Only the last 5 items listed under documentation (summary of the ethical analysis; identification of the ethically appropriate decision maker(s); options considered, and whether they were deemed ethically justifiable; explanation of whether agreement was reached; recommendations and action plan(s)) deal with ethical analysis to have any form of ethical content. (Core Competencies, 19)

While there is much guidance on the process of ethics consultation, and mention is made of ethical standards or ethical analysis, the document does not articulate exactly what standards are to be applied (except the procedural standards), or what ethical analysis looks like. Emphasis is placed on technique rather than to contentful discussions of the goods of medicine, or the deep and robust understanding the metaphysical moral commitments of patients. Process and interpersonal skills, along with managerial skills are the skills most needed to run the HCEC program.

Skills of ethical assessment are also required, focusing on the consultant's ability to:

- discern and gather relevant data (e.g., clinical, psychosocial)
- assess the social and interpersonal dynamics of the consultation (e.g., power relations, racial, ethnic, cultural, and religious differences)
- distinguish the ethical dimensions of the consultation from other, often overlapping, dimensions (e.g., legal, institutional, medical)
- articulate the ethical concern and the central ethics question
- identify various assumptions that involved parties bring to the consultation (e.g., regarding quality of life, risk taking, institutional interest, unarticulated agendas)
- identify relevant values of involved parties
- identify the consultant's own relevant moral values and intuitions and how these might influence the process or analysis (Core Competencies, 23)

The skills of analysis that the CE consultant must possess are the ability to:

- access the relevant ethics knowledge (e.g., health care ethics, law, institutional policy, professional codes, research/scholarship, and religious teachings)
- clarify relevant ethical concepts (e.g., confidentiality, privacy, informed consent, best interest, professional duties, etc.)
- critically evaluate and use relevant knowledge of health care ethics, law (without giving legal advice), institutional policy (e.g., guidelines on withdrawing or withholding life- sustaining treatment), and professional codes in the consultation.

To critically evaluate and use relevant knowledge, the consultant must also have the ability to:

- apply relevant ethical considerations in helping to analyze the consultation identify and justify a range of ethically acceptable options and their consequences
- evaluate evidence and arguments for and against different options
- research peer-reviewed clinical and bioethics journals and books, and access relevant policies, laws and reports, using the Internet and/or libraries
- recognize and acknowledge personal limitations and possible areas of conflict between personal moral views and one's role in ethics consultation (e.g.,

9 On the Social Construction of Health Care Ethics Consultation 185

accepting group decisions with which one disagrees, but which are ethically acceptable)
• be familiar with diversity among patients, staff and institutions and address it in relation to an ethics consultation (Core Competencies, 23–24)

Thus, there seems to be a commitment to a neutral and supposedly contentless process. For the most part, one would think that little harm can come to the patients and families who are the beneficiaries of HCE consultation. Yet, there is a content that is imported, namely the content of a bureaucratic society.

Thus, as opposed to the falsehoods of the extreme "authoritarian" or "pure consensus" models, and as opposed to professional, medical, legal or spiritual advice given by practitioners of practices of nursing, medicine, law, and pastoral care, the ethics facilitation model of HCEC is founded upon a "generally agreed-upon" standard, which, because it is generally accepted by HCE consultants, operates as the acceptable and true model for clinical ethics consultation. After all, who could be in favor of authoritarian models or mindless consensus models? And since HCEC cannot compete with the professions of nursing, medicine, law, and pastoral theology, and since HCEC cannot articulate their own goods, they must establish a set of merely bureaucratic guidelines that purport not to import robust metaphysical moral values. And with a slight of hand, they proclaim by fiat a set of thin procedures, all the while deploying a gaze.

9.4 Constituted Gazes

While the authors of the revised Core Competencies attempt to establish the importance of ethical knowledge and moral reasoning, this section on ethics, philosophical and religious, only receives a bulleted paragraph, about one-fifth of a page. (Core Competencies, 28) Yet, it is not the lack of attention to the various schemes of philosophical and religious ethics that is problematic. It is instead the deployment of the process itself that comes to constitute what gets labeled as the ethical, and this is best seen in the emphasis placed upon the assessment of the processes of HCEC. Occupying a full third of the 53 pages of the Core Competencies draft document is a section devoted to the evaluation of HCEC. The ethics of HCEC has become quality improvement.

In subservience to the efficiency and effectiveness paradigm, the Core Competencies have bound themselves to the Quality Improvement movement. Again, the term "quality" carries with is some sort of good, though it is not clear what that good is. Human resource and quality improvement scholars utilize the tools of social science to define and measure qualities. Those doing research with social scientific tools realize that social objects cannot be read in exactly the same way as natural objects, such as rocks. So an elaborate process is created to assure that social scientists, including quality improvement researchers, are speaking about the same things in the same way. (Babbie, 2004, p. 119) The "quality" portion of "quality improvement," is just such an "object" socially constructed in order to make it measurable.

As in all social scientific research, the first step is to create a definition of what it is that one wishes to know. (Babbie, 2004) This research relies heavily on the construction of definitions that refer to elusive social-practical realities. Yet, conceptual definition is not enough, for the researcher must also define a set of indicators that will allow the researcher to recognize and register the concept as something that is empirically observable. For example, Swiderski et al. (2010) want to propose that their quality instrument is designed to pick out those things that are salient to measuring a "quality ethics consultation." In creating an operational definition, say "quality ethics consultation," the social scientist can see "quality ethics consultation" in operation. In this way, then, "[c]onceptualization is the refinement and specification of abstract concepts, and operationalization is the development of specific research procedures (operations) that will result in empirical observations representing those concepts in the real world." (Babbie, 2004, p. 132) Thus, there is a circularity of the process that even social scientists recognize: define, operationalize, test, redefine. Thus, the definition of "quality ethics consultation" articulated by Swiderski et al. is run through what social scientists call the "conceptual funnel" (Babbie, 2004, p. 131; Fisher, 2004, pp. 130–131) so that a more narrow and delimited definition emerges on the other side of the funneling process.

What is a "funneling process" to a social scientist can also be seen as an epistemological circuit. "There arises. . . an epistemological circuit whereby knowledge is based entirely on objects, whose 'being' does not exceed the extent to which they are known." (Pickstock, 1998, p. 63) Now regarding research generating quality ethics consultation instruments, "what is measureable becomes the standard for what is "knowable," which in turn becomes the standard for what "is."" (Bishop, 2009, p. 342) This epistemological circuit is precisely what is at work in those attempts to define quality HCEC.

This epistemological circuit shows at least three things about the proposals of Swiderski et al. First, there is an assumption that the quality improvement assessment assists the clinical ethics consultant by systematically evaluating his procedures and standards. The basis for that assumptions is that the circularity of the process leads to better definitions, more narrow definitions, definitions that pick out certain relevant features of the clinical ethics consult. The second and most important aspect of this "funneling process" or "epistemological circuit" shows that as these definitions ossify into standards of practice, they become more and more the ready-made rules of engagement. The third aspect of the process, then, is that the circularity of the process would come to structure the practice of the expert. The expert designs the instrument; and the instrument assures the expert that he is in fact expert.

Before setting out to operationalize a metric that detects the presence or absence of a quality ethics consult, Swiderski et al. (2010) set out to define quality clinical ethics consultation. Then, they can be certain that their inventories and evaluation mechanism matches their definition. Oddly enough, and contrary to the claims of the Core Competencies regarding a supposed emerging concensus, Swiderski et al. note

9 On the Social Construction of Health Care Ethics Consultation

that there is no consensus among the panel of experts that they gathered together. Still, even without a consensus definition, they persist in believing that their operationalized definitions are getting at some sort of accurate way of analyzing the processes of HCEC. For unclear reasons, they hold their instrument in high regard. For social scientific research, the success of any QI tool should be dependent on the correspondence between a determinate concept of quality (an accurate definition) and its applicability in observations of actual practice. Yet, as evidence of the "success" of the QI tool, Swiderski et al. point to how the QI instrument began to be used:

- The consultant who objected most strenuously to the QI tool described in an exit interview how her practice had changed because of it: it reminded her to include certain information, she organized her notes according to its format, and she began adding "a line or two" of educational content to each note.
- A senior consultant on a very busy service also felt that it helped to strengthen her documentation in her notes; she also used the QI tool to approach the QI Committee in her hospital to develop a formal QI mechanism for her service.
- A junior consultant began carrying the QI tool with him as a vital "checklist" when performing consults.
- A group of consultants who operated within a network of ambulatory sites and began the project with no written chart notes used the tool as a template to begin developing a procedure for formal notes. (Swiderski et al., 2010, p. 71)

The tool began to be used to inform practice; and its use in informing practice is evidence that it is useful. Never mind that the usual "funneling process" was not followed. In other words, even before the tool could establish a correspondence between a determinate conception of quality and observations of actual ethics consultations it was found to be useful, concluding that the instrument should be used. The truth of the QI tool can be found in the fact that it usefully changed behavior.

The fact that "almost all participants reported a positive effect of using the QI tool on their practices" gives evidence that the tool is the right tool and that it measures the right things. (Swiderski et al., 2010, p. 72) Of course, the problem with their instrument is not that they did not follow the standard approach to designing such instruments, because the truth is that the best designed instruments only achieve more power to control. (Bishop et al., 2010, p. 82) In this case, the tool began to shape, not only what the clinical ethicist does, but how the the ethical process and those patients and families that are put through the process now must conform to the truth deployed by this instrument.

I am sure that good intentions motivated the construction of this QI tool, but the circularity of all such instruments of assessment cannot be ignored. As noted by was developed with all good intentions. The ready-made rules of Swiderski et al.'s to create the practitioners in its own image, or rather in the image that the nebulous group of experts have defined. In other words, the fact that both a senior and a junior faculty-member began to mold their practice to the instrument suggests that a certain clinical ethical gaze is deployed. As I, along with colleagues have noted elsewhere, there are repercussions to the development and deployment of such instruments of standardization:

1) this nascent QI instrument, which has already taken on a life of its own, will ossify and will be deployed as a new normative gaze by which the clinical ethicists encounters the body of the patient.
2) by more firmly establishing the gaze through standardization and through instruments designed to control quality, the circuit will more firmly establish the clinical ethicist as an expert because he deploys ready-made expertise. (Bishop et al., 2010, p. 82)

Every tool of assessment defines not only its object, but those who do the assessing. (Bishop, 2009)

The same holds for the Core Competencies document where 15 pages are devoted to quality assurance and improvement, and one-fifth of a page is devoted to ethics and ethical content.

For each component of health care ethics consultation services that can be evaluated, we provide a definition of the component and how it applies to ethics consultation, provide examples (where available) of available empirical data on ethics consultation relating to that component and tools that have been used to evaluate that component, and make recommendations for evaluating and improving ethics consultation. (Core Competencies, 37)

Thus, the stage is set for defining the goals HCEC, not in the terms the goods and purposes of a practice, but in terms of quality improvement and the function of the processes. Their justification will be found in the intertwining narratives of health care institutions and medical practice adopting their goods as their own: medicine's good is an evidence base; the hospital's good is quality improvement, quantified for purposes of making them seem more real. These quantitated qualities will be deployed without an ounce of understanding of the oxymoronic nature of quantitated qualities. (Bishop et al., p. 285)

Thus, we can return to the physician HCE consultant who stood at the back of the ASBH conference bemoaning the unethical behaviors in small hospitals all over the US that do not have HCEC at their facilities; an yet HCE consultants consistently point out that they are not moral or ethical experts. So perhaps we can better understand what the framers of the Core Competencies mean. This sentiment that HCE consultants do not bring medical, legal, spiritual, or even moral expertise is in one sense true. They have hidden any moral or ethical expertise in one-fifth of a page, and developed 50+ pages of process driven ethics, subservient to the goods of the bureaucracy of medicine. The goods they deploy are the goods of quality, but of what exactly quality is predicated is not at all clear. Thus, just as Alasdair MacIntyre has noted, the highest goods are the goods of efficiency and effectiveness. The ethicist then is a bureaucratic manager, guiding a process toward unknown and interchangeable ends and goods. (MacIntyre, 1984, pp. 26–27, 30–32, 74–78, 85–87) In HCEC, the supreme good is efficiency and effectiveness of the process, the intermediary goals not the goods of a valued practice.

References

Babbie, E. 2004. *The practice of social research*, 10th ed. Belmont, CA: Wadsworth/Thomason.

Bishop, J.P. 2009. Biopsychosociospiritual medicine and other political schemes. *Christian Bioethics* 15: 254–276.

Bishop, J.P., J.B. Fanning, and M.J. Bliton. 2009. Of goals and goods and floundering about: A dissenseus report on clinical ethics consultation. *HEC Forum* 21(3): 275–291.

Bishop, J.P., J.B. Fanning, and M.J. Bliton. 2010. Echo calling Narcissus: What exceeds the gaze of clinical ethics consultation? *HEC Forum* 22(2): 73–84.

Callahan, D. 2000. *Troubled dream of life: In search of a peaceful death*. Washington, DC: Georgetown University Press.

Engelhardt, H.T., Jr. 1991. *Bioethics and secular humanism*. London: SCM.

Engelhardt, H.T., Jr. 1996. *The foundations of bioethics, 2nd ed*. New York: Oxford University Press.

Fisher, P. 2004. The importance of variable names. In *The practice of social research*,10th ed., ed. E. Babbie, 130–131. Belmont, CA: Wadsworth/Thomason.

Foucault, M. 1994. *The birth of the clinic: An archaeology of medical perception* (trans: Sheridan Smith, A.M). New York: Vintage Books.

Fox, E., K.A. Berkowitz, B.L. Chanko, and T. Powell. 2006. *Ethics consultation: Responding to ethics questions in health care*. Washington, DC: Veterans Health Administration.

Frolic, A., and P. Chidwick. 2010. A pilot qualitiative study of "conflicts of interests and/or conflicting interests" among Canadian bioethicists. Part 2: Defining and managing conflicts. *HEC Forum* 22(1): 19–29.

Jonsen, A. 1998. *The birth of bioethics*. New York: Oxford University Press.

MacIntyre, A. 1984. *After virtue*, 2nd ed. Notre Dame, IN: University of Notre Dame Press.

McCullough, L.B. 1998. *John Gregory and the invention of professional medical ethics and the profession of medicine*. Dordrecht: Kluwer.

Pickstock, C. 1998. *After writing: The liturgical consummation of philosophy*. Oxford: Blackwell.

Swiderski, D.M., K.M. Ettinget, M. Webber, and N.N. Dubler. 2010. The clinical ethics credentialing project: preliminary notes from a pilot project to establish quality measures for ethics consultation. *HEC Forum* 22(1): 65–72.

Index

A

AAUP code, 134
Abel i Fabre, F., 167
Academic and clinical dimensions, 8
Activism as a function of bioethicists, 133
Adversarial ethics, 164
After God, 5–6, 21
Agnosticism, 6
Alexander, S., 32
Alora, A.T., 137, 170
AMA code, 134
American Academy of Medical Colleges, 128
American Association of Bioethics, 125
American bioethics, hegemony of, 170
American Correctional Association, 38
American culture of 1960s and 1970s, 13
American dream, 51
American Society for Bioethics and
   Humanities (ASBH), 20, 115, 117,
   123, 128, 165, 178, 181
Analytic skills, 8, 163
Andorno, R., 111, 118
Anti-elitist sentiments, 152
Anti-paternalism, ideology of, 12, 91
Anti-traditional trends, 17
Anti-Vietnam war protests, 9
Antommaria, A.H.M., 127–128, 138
Arafat, Yasser, 59
Arnold, M., 132, 169
Arnold, R., 132
Assessment tool, 188
Atheism, 6, 41, 158, 170
Attraction or revulsion, feelings, 91
Ausilio, M.P., 146
Authoritarian clinical ethicist, 182
"Authoritarian" or "pure consensus" models,
   185

Autonomy, 12–14, 41, 51–52, 56, 74–75,
   82–83, 85–95, 100–101, 117–118,
   142–143, 154, 160, 180
 bioethical principlism, 89
 respect of patients, 89
 justice, and beneficence, 41
 paradigm of, 10
 understanding of, 4, 12–13, 86, 92–93
Autonomy movement, 51

B

Babbie, E., 185–186
Babydom, 55, 64
Bacon, Francis, 78
Baker, R., 26, 76, 169
Baptists, 136
Bartlett, B., 61
Basic human rights, 101–103, 108, 112–114,
   119–120
Baumeister, R.F., 51
Bayertz, K., 170
Beauchamp, T.L., 3, 12–13, 34, 44, 51, 73, 82,
   85–93, 95, 142–143
Beckwith, F.J., 119
Beecher, H.K., 32
Belmont Report, 8, 35, 43–46
 corresponding guidelines, 44
 de facto federal policy, 44
 ethical analysis of human research, 45
 general principles, 44
 investigations and analysis, 44
 legal analysis, 45
 principles and guidelines, 43
 risk-benefit assessment, 44
 spirit of Belmont, 44
 task, 43
Beneficence, non-maleficence, and justice, 85,
   89
Bible reading in school, 153

H.T. Engelhardt, Jr. (ed.), *Bioethics Critically Reconsidered,*
Philosophy and Medicine 100, DOI 10.1007/978-94-007-2244-6,
© Springer Science+Business Media B.V. 2012

Index

Bill Clinton's National Bioethics Advisory Commission, 4, 170
Bioethical pluralism, 3, 14, 16, 18, 160, 166
Bioethical principles, 117–118
Bioethical principlism, 12, 85–87, 89, 91–93, 95
Bioethical professionalism, 20–21, 175
"Bioethicist of the month" feature, 49
Bioethicists
  activism as a function of, 133
  functional diversity of, 134
  multi-tasking of, 165
  obligation of, 133
  potential function of, 133
  politics, connection, 4
Bioethics enterprise, 129–144
  disciplinary differences, 129–131
  functional diversity, 132–135
  religious, cultural and moral pluralism, 136–139
  sub-fields/sub-specialization, 135
Birth of bioethics, 87
Bishop, J., 20–21, 115, 177–178, 186–188
Blair, H., 154
Boba, R., 116
Bole, T.J., 115
Bonnar, A., 168
Bosk, C., 133
Boskin, M.J., 66
Bourgeois socialism, 62
Bouscaren, T.L., 168
Boyle, J., 141–143
Brock, D., 124
Brody, B., 32, 132
Brooks, A.C., 58
Brown, N.O., 54, 64
Brunekreef, B., 116
Buchanan, A., 86–87, 94, 108–109, 124
Buchanan, J., 106
Buckley, M., 10, 25
Burden of proof, shifting, 110–111
Burkemper, J., 145
Burkhauser, R.V., 61
Bush, George, 58

C

Callahan, D., 115, 132, 177
Canonical morality, 5, 7, 14–15, 24–25, 156–157
Capellmann, C.F.N., 168
Caplan, A., 56–57, 60
Card, D., 67
Cardinal cultural normativity, 162

Caring concern for moral issues, 165
Carpenter, A., 16, 123
Carrington, W.J., 62
Cartwright, S.A., 169
Cassell, E.J., 44
Castro, R., 169
Catholic Health Association, 168
20th Century liberalism, 62
Chambers, T., 129
Character of bioethics, 15, 20, 93
Chattopadhyay, S., 137
Chavez, H., 59
Chen, Xiaoyang, 137
Cherry, M.J., 15, 99, 113, 118–119
Chidwick, P., 178
Children involvement in research, 40–43
  customary taxonomic approach, 40
  formal reasoning, 42
  minor-increase-over-minimal-risk, 42
  nontherapeutic research, 42
  parental authority, 40
Children privacy, restrictions, 110
Childress, J., 13, 51, 85–86, 88–95, 130, 142–143
Childs, B.H., 115
Cho, M.K., 132
Christian bioethicists, 136
Christian polity, 153
Christians, non-denominational, 136
Chrysostom, St. John, 24–25
Civil rights movement, 9, 50, 55, 58, 155
Civil rights of patients, advocacy, 12
Clark, P.A., 104
Classical antiquity, paganism, 154
Classical metaphysics, collapse, 159
Classroom-based smoking abstinence, 57
Clement of Alexandria, 24
Clinical ethicist
  hiring a, 22
  role
    attorneys, 165
    ethics experts, 165
    legal advisers, 165
    risk managers, 165
Clinical ethics, 1–3, 5, 15–17, 19, 21, 23, 115, 130, 132, 134–136, 139, 141, 152, 164, 178–179, 182–183, 185–186
Clinical ethics consultants, 1, 178–179
  correct model for, 182
  practice of, 180
Cloning, 14, 99, 113, 164, 170
Clouser, K.D., 72–73
Code of ethics, 16, 123, 131, 134, 168

Index

Codronchi, G., 169
Common morality theory, 143
Complex social phenomenon, 151–153
Conceptual funnel, 186
Consequentialists, 2
Consultation process, 183
Contemporary culture war view, 119
Content and scope of bioethics education, 137
Contraceptive pills, placebo trial of, 32
Convention on Civil and Political Rights, 107
Convention on Economic, Social, and Cultural Rights, 107
Convention on the Rights of the Child, 15, 101, 104, 107, 109–112
  censor information, 102
  education of the child, 102
  free and compulsory education, 104
  freedom of association and peaceable assembly, 103
  freedom of thought, conscience, and religion, 102
  homosexual acts, consent for, 102
  pediatric decision-making, 102
  rights to privacy, 103
  rights to sexual pleasure, 102
  sex education, 102
Conveyance of authority, 14
Conveyance of permission, 14
Cook, R.J., 103
Coppens, C., 168
Copray, N., 169
Core ambiguities, 21
Core competencies, 16–17, 20–21, 115, 165, 181–186, 188
Counter-cultural movement, 62
Critical evaluation, 184
Cultural centrality, 162
Cultural commitments, 136
Cultural-moral vacuum, 153–156
Cultural norm, non-acceptance of, 137
Cultural revolution in China, 9
Cultural revolution of the 1960s, 50–51
Cultural struggle, 154
Cultural transmission, 50
Culver, C., 86, 88, 94
Curtius, E.R., 169

D

Daniel-Rops, H., 168
Daniels, N., 53–54, 64, 66, 108, 118
Decisions of patients, respect, 93
Dellinger, D., 59
de Melo-Martín, I., 132

Democratic party, politics, 60
Deprofessionalization of medicine, 17, 156
Deprofessionalized medical ethics, 71–74, 82–83
Deprofessonalizing bioethics, 83
Derse, A.R., 129
Descartes, R., 25, 159
DeVries, R., 132–133
Differentiation process, 55
Dionysian religion, 56
Disciplinary and functional diversity, 140, 144
Discipline's multiplicity, 2
Disclosure, 2, 5, 13, 87
Dispute mediation, 23, 166
Dispute mediators, 19, 153, 165
Dockery, D.W., 116
Dresler, C., 107
DuBois, J., 145
Dupré, L., 159
Dworkin, G., 94

E

Early Enlightenment, 76
Eberstadt, N., 66
Ecstatic sex, aftermath of, 55
Edelstein, L., 12
Ego, 54
Eichendorff, J.F. von, 168
Elliott, C., 133
Ellis, A., 169
Embryonic stem cells, use of, 164
The Encyclopedia of Bioethics, 72–73, 82, 87–88, 151
Engel, G.L., 79
Engelhardt, H.T., Jr., 1, 5–6, 14–19, 23, 25, 87, 100, 112–113, 115, 124, 132, 137, 143, 151, 161, 177
Enlightenment, 8, 11, 23, 111, 113, 126, 154, 158–161
Enlightenment aspiration, 11, 160
Enlightenment concept, 111
Enlightenment expectations, 23
Episcopalians, 136
Epistemological circuit, *see* Funneling process
Epstein, R., 119
Erotic love, 55
Established religion, 153, 167
Ethical advice, 22
Ethical assessment, skills, 184
Ethical principles, 31, 33, 43–44
Ethical theory, 71, 73–74

194 Index

Ethics consultants, 1–2, 5, 17–23, 151–153, 165, 178, 181
   aka consultants, 1
   core competencies, 165
   disclosure, 5
   as dispute mediators, 153
   guidance, 2
   as legal liability risk manager, 153
   role of, 23, 165
   services, 18, 152
Ethics consultation, 2, 19, 21, 97, 132, 152, 165, 177–189
Ethics guidance, 2
Evidence-based medicine, 79
Expert consensus, 124
Eye-sight quickness, 80

F
Fallick, B.C., 62
False claims and false expectations, 5
Family-based healthcare savings accounts, 109
Fan, R., 137, 166, 170
Fetal Research, 33, 35–36
   doctrinal differences, 36
   fetal experimentation, 36
   fetus ex utero, 35
   substantial consensus, 36
Ficarra, B.J., 168
Finegan, T.A., 61
Fink, C.T., 132, 168
Finney, P.A., 168
Finnis, J., 141–143
Fisher, P., 186
Fiske, E., 132
Flexner, A., 162, 169
Flood, P., 168
Food and Drug Administration (FDA), 38
Foucault, M., 182
Foundational disagreement, 3
Foundations of morality, 164
Fox, E., 183
Fox, R.C., 14–15, 99–100, 113, 115, 127
Framework assumptions, 125
Framework for identity formation, 54, 59, 64
Frank, J.P., 169
Free-markets, 113–114
French Revolution, 154
   anti-religious violence, 168
Friedman, T.L., 58
Frolic, A., 178
Fuchs, V.R., 61, 67
Functional capacities, 53
Funneling process, 186

G
Gallup Polls, 140
Gauderman, W.J., 116
Gay, P., 154
Gbadegesin, S., 137
Genetic engineering, 32
Geraghty, K., 133
German secularization, 154
Gert, B., 86, 88, 94
Gitlin, T., 59
Golden rule, 7
Goldwater, Barry, 9
Goodman, W., 32
Grattan, C.H., 169
Gregory, G., 11, 76, 78–83, 132, 169
Gregory, J., 11, 76, 132
Grey zone, 19, 22, 153, 164–165
Griese, O.N., 168
Grisez, G., 141–143
Guthrie, W.K.C., 24

H
Hayes, E., 168
Health Care Ethics Consultants (HCEC), 1, 178
   goals of, 181–182
Health policy theory, 52
Health welfare entitlements, 104–106
   consensual interaction, 105
   governmental duties, 105
Healy, E.F., 168
Hegel, G.W.F., 159, 169
Hellegers, A., 135, 160, 167
Hermeneutic circles, 160
Hessler, K., 106
Heteronomy, 86
   See also Autonomy
High Middle Ages, 160–161
Hinkley, A.E., 8, 12–13, 85
Hippie legacy, 11, 50
Hippocrates, 77, 81, 168
Hippocratic ethos, 12
Hippocratic medical ethics, 82
Hippocratic medicine, 132
Hippocratic Oath, 12, 82, 168
History of bioethics, 8–14
Hoeveler, J.D., Jr., 169
Hoffman, A., 50
Hoffmann, F., 78
Hofmann, H.H., 168
Holy Roman Empire, 154
Homeschooling parents, 123
Homosexuality, 164

Index 195

Hoshino, K., 170
Hottois, G., 169
Human embryos, 124
Human experimentation, 31–32,
43–45
ethical conduct, 43–44
justification, 43
principles and rules to regulate, 45
Human immunodeficiency virus (HIV), 56,
103–104
Human rights, political announcements
of, 106
Hume, D., 78, 154, 159
Hunter, J.D., 14
Huntington, S.P., 167, 170
Hussein, Saddam, 58
Hypocrisy of parents, 50

I
Identity deficit crisis, 51
Identity formation
components, 51–52
problem of, 55
Ideological agendas, 17, 152
Ideological and political pluralism, 16
Ideologically partisan character, 170
Iltis, A.S., 16, 123, 131
Immanent domestication, 158–160
Incapacitated patient, 124
Individualism, 51–52, 55, 59–60, 62, 65, 85,
94
vs. collectivity, 60, 62
hippie synthesis, 55
older forms of, 60
vs. sociality, 55
Industrial revolution, workplaces of, 77
Inherent coercion, 39
Intellectual undertakings, hybrid of, 167
Intra-bioethical disagreement, 14
Irregulars, 76

J
Jackson, S., 153
Jaggar, A.M., 107, 117
Jahr, F., 167
Jewish bioethicists, 136
Jewish Chronic Disease Hospital, 32
Joachim, J.M., 116, 120
Jonsen, A.R., 71, 73, 78, 87, 127, 132, 145,
167, 177
Josephism, 154
Jouanna, J., 82

K
Kant, I., 5, 10, 12–13, 25, 86, 88, 90–93, 95,
110, 159
Kantian autonomy, 86, 90–93
inclination-driven choices of patients, 93
moral agents, 90
moral rationality, 90
principled priority, 92
Kantian physician, 13–14, 92–93
Kass, L., 4
Kawachi, I., 66
Kelly, G.S.J., 132, 168
Kelly, P.J., 57
Kennedy Institute, 3, 10, 23, 151, 160–161
Kennedy, Ted, 32
Kenny, J.P., 168
Khushf, G., 167
Kimball, R., 55, 64
Kipnis, K., 115, 134
Klein, E.P., 167
Konold, D.E., 169
Kopelman, L.M., 125, 129–130, 135, 145
Korff, W., 169
Kosmin, B., 140
Krause, E.A., 155
Krueger, A.B., 67
Kuehn, M., 25
Kumar, N.K., 137
Kymlicka, W., 104, 117

L
Laertius, D., 24
Lantos, J.D., 134
La Rochelle, S.A., 132, 168
Leary, Timothy, 50, 55
Lebow, D., 66
Lecaldano, E., 169
Legal or social sanctions, 22
Leibnitz, G., 25, 159
Leone, S., 169
Lepenies, W., 169
Levi, D.L., 51
Liability risk management, 19, 153, 166
Li, En-Chang, 137
Life-plan
contingencies, 54
inventory, 53
Lifestyle pluralism, 60
Life-taking diseases, 75
Little, D., 75, 110
Little peculiarities, 76
Locke, J., 118
Loewy, E., 61–62, 128

Lord's Prayer, recitation, 153
Lumitao, J.M., 137, 170
Lutherans, 136
Lyotard, J.-F., 160

M
MacIntyre, A., 21, 81, 179, 188
Macro-level bioethics, 4–5
Mahoney, M.J., 35
Managerial skills, 184
Market-based health care, 114
Marshall, M.F., 133, 145
Martin Luther King, Jr., 50
Matthews, M., 66
Matusow, A.J., 10, 54–55, 63–64
McCarthy, J., 51
McCullough, L.B., 4, 8, 11–12, 26, 71, 76, 78–82, 85, 132, 177
McElhinney, T., 169
McFadden, C.J., 168
McGovern, George, 9
Meadowcroft, J., 118
Medical ethics, long tradition, 73, 76
"Medical factor", 75
Medical-moral commitments, 155
Medical paternalism, 4, 11–13, 26, 73, 79, 82–83, 85–88, 92–94
    and autonomy, 86–88
        anti-trust exemption, 87
        decision-making, 86
        guild status of medicine, 87
        reasonable-and-prudent-person standard, 87
        violations of moral rules, 88
    critics, 26, 93–94
    crusade against, 85, 87
    definition of, 94
    Kantian autonomy, 91–92
    necessary condition for, 83
    special evil, 11
    socio-political agenda, 4
Medieval Western cultural-philosophical synthesis, 158
Messikomer, C.M., 120
Metaphysical pluralism, 160
Micro-level bioethics, 4–5
Miller, F.G., 134
Minnesota Human Rights Education project, 117
Missa, J.-N., 169
Mitford, J., 36
Mondale, W., 32
Montessori teachers, 123

Montgomery, K., 129
Moral/ideological commitments, 136
Moral and epistemological ambiguity, 106–107
Moral anthropology, 111–113
    religious and cultural assumptions, fragmentation, 112
    moral cosmopolitanism, 112
    moralizing pronouncements, 112
    traditional cultures and religions, 112
Moral commitments, 2, 4, 8, 10, 17, 19, 55, 114, 137, 164–165, 184
Moral consensus, 4, 56, 107–109, 113
Moral controversies, 3, 113
Moral-cultural rigidity, 50
Moral disagreements, 4, 22
Moral disorientation, 18, 162
Moral epistemology, 9, 155, 168
Moralities
    philosophical accounts, 157
    pluralism of, 14
    general rules, 74
    and rationality, 86, 161
Morality status, reassessing, 6
Moral-philosophical project, 24, 158
Moral pluralism, 1, 3, 6–7, 9, 14–15, 17–18, 20–22, 24, 60, 109, 152, 157–161, 164, 170
    challenges of, 157–158
    disputes, 15
    deflation of authority, 24
    intractable, 1–3, 17–18, 21, 158
    emergence of post-modernity, 161
    extent of depth of, 18
    reality of, 109
    secular, 20
Moral-theoretical and meta-ethical issues, 2
Moral values and life-style choices, 101
Morgan, S., 116
Multiculturalism, 126
Multi-tasking, 165

N
Nagel, T., 53
Napoleon, 154
National Commission for the Protection of Human Subjects, 8, 33, 35, 37–38, 40–41, 43, 45
National Research Act, 33
Nature of bioethics, 1, 9
Nazism, 128
Neumark, D., 61, 67
Neuroethics, 135
Neutral bioethics, 137

Index 197

Neutral theory of the good, 53
New Age religion, 56
New Age spirituality, 56
Nietzschean, 113
Noble moral crusaders, 12
Non-belief in a particular religious principle, 137
Non-morally-normative services, 163
Nuremberg Code, 37, 44
Nuremberg trials, 32
Nursing ethics, 135
Nutton, V., 82

O

O'Donnell, T.J., 168
Oken, D., 11
O'Leary Morgan, K., 116
Oleksa, M., 153
O'Neill, W.L., 63
Organ transplantation, 32, 124, 132
Organizational ethics, 132, 135–136
Organized medicine, "corporation spirit", 80
Orthodox Christians, 108, 136
Orthodoxy, 18, 164

P

Parker, L., 133
Parsi, K., 133
"Partner changing", 56
Paternalism, 4, 11, 26, 73–74, 79, 81–83, 85–89, 91–95
   animus against, 26
   and autonomy, 86–88
   criticism of, 26, 85, 93
   definition, 87–88
   Kantian autonomy, 86, 91–92
   necessary condition, 83
   special evil of, 11
Paternalistic physician, 12, 77
Patient-physician relationship, 81
Pellegrino, E.D., 125–126, 169
Percival, T., 11, 76, 79–80, 132
Permanent vegetative state, 75
Personal identity, meta-criterion, 53
Pessini, L., 137
Physician-critical medical morality, 10
Physician paternalism, 51
Physician-patient relationship, 5, 12, 72, 81, 87
Physician's status as guides, 152
Pickstock, C., 186
Pleomorphism, 7
Political ideology, 15, 49, 60, 99–115
Political liberalism, 126

Porter, D., 76
Porter, R., 76
Postmodernism, 126
Potter, V.R., 135, 137
Practice of Bioethics and Clinical Ethics Consultation, 14–19
Practitioners, 125
Predatory power of practitioners, 83
President's Council on Bioethics, 4, 170
Principle of autonomy, 12–13, 85–86, 89–90, 93, 143
Principle of respect for autonomy, 74, 89
Principles of Biomedical Ethics, 3, 12, 44, 85, 88, 90
Principlists and casuists, 41
Prisoners, involvement in research, 36–40
   drug testing, 36, 39
   informed consent, 39
   nature and extent of research, 37
   nontherapeutic research, 37
   phase 1 drug testing, 37–38
   risks of testing, 38
   staff and peer pressure, 37
Private-practice physicians, 77
Privitera, S., 169
Professional competency, evaluation of, 20
Professional integrity, 17, 75, 81
Professional medical ethics, 71–77, 79–83, 155
   centrality of, 74
   conservative defense, 81–83
   invention of, 76–81
   promise-making and keeping, 75
   scientific and clinical competence, 79
Progressivist agenda, 14
Promiscuity and prostitution, 56
Protagoras, 24
Protestant Christianity, 167
Protestantly-oriented Bible, 153
Protestant polity, 153
Public Health Service, 32
Public ideology, 100–106
   deconstruction of family, 100
   human rights, 100–104
   responsibilities and duties of parents, 101
Public school teachers, 123
Pure consensus approach, 182
Pure consensus clinical ethicist, 182

Q

QI tool, 187
Qiu, Ren-Zong, 170
Quality ethics consultation, 186
Quality improvement assessment, 21, 186

Quality improvement (QI), 21
Quasi-casuistic practice, 9
Quasi-empirical considerations, 157
Quasi-precedential practice, 9
Quinlan, J., 74

R
Radical individualism, hippie synthesis of, 55
Rage politics, 58–62
    language of moral revulsion, 61
    lifestyle pluralism, 60
    moral disapprobation, 61
    moral pluralism, 60
    neoconservatism, rise of, 58
    transition to political activism, 59
RAND Corporation, 33
Rape, 58
Rasmussen, L., 124, 128, 130, 133, 137, 145, 170
Rational action theories, 113
Rau, W.T., 169
Rawls, J., 52–54, 64, 143, 157–158, 169
Rawls's theory, 53
Reading, R., 101, 111
Reassessment of bioethics, 1–24
Rector, R.E., 66–67
Reformist bioethics, 81
Reich, W.T., 72, 145, 135, 167
Religion of erotic love, 55
Religious
    artifact, 102
    commitments, 136–137, 141, 144
    differences, 100
    ethics, 135
    faith and cultural life, 109
    goals, 104
    liberty, 111
    morality, 110
    moral teachings, 57
    orthodoxy, 111, 164
    right to religious freedom, 153
    traditional group, 57
Renaissance, 158
Renal dialysis, 31
Research ethics, 132, 135
Rie, Michael A., 23
Ringheim, K., 101, 115–116
Risk management, 5, 23, 164
    concerns for, 18
    moral disorientation, 18
Risse, G.B., 77–78
Roman Catholic, 19, 25, 108, 131–132, 136, 160–161, 165, 168

Roman Catholic health care institution, 131
Roman Catholicism, 25, 160, 168
Roman Catholic manualist tradition, 132, 168
Romanides, J., 25
Rothenberg, L.S., 146
Rothman, D.J., 31–32, 80, 132, 167
Royce, J., 52–53, 65
Rudd, J., 66
Rüdiger, H., 169
Russo, G., 169
Ryan, K., 33

S
Salient moral and metaphysical pluralism, 156–158
Same-gender marriage, Proponents of, 119
Samet, J.M., 116
Sandel, M.J., 64
Sandström, T., 116
Sanford, A., 168
Sass, H-M., 170
Schaefer, J.B.P., 169
Schneewind, J.B., 154
Schwartz, L.R., 57
Secular fundamentalism, 106
Secularism, 25, 126, 152
Secularization of 1803, 154
Secularization of expectations, 155
Secularization processes, 152, 156
Secular morality, 5–6, 18, 21, 24, 110, 113, 120, 160
Self-alienation, 10
Self-determination, 10–11, 100–101
Self-directed moral fashion, 86
Self-physicking, 76, 83
Self-realization, 10, 13, 52, 58, 101
Sensual experiences, 54
Sensual indulgence, 51–52
Sex education, 57, 64, 102
Sextus Empiricus, 24
Sexual intercourse, satisfying, 57
Sexual manipulation, 58
Sexual morals of American youth, 57
"Sex workers", 56
Shriver, S., 135, 160, 167
Simon, A., 137
"S" in bioethics, 139–144
    common morality approach, 143
    disciplinary diversity, 140
    functional diversity, 140–141
    pluralism and diversity, 139
    religious, cultural and ideological pluralism, 140

Index

religious diversity, 141
shared morality, false declarations, 144
sub-specialization, 141
superiority of a moral position, 143
Sittlichkeit, 22
Skills of analysis, 184
Snow, C.P., 162, 169
Social constraints, 54
Social construction of health care, 177, 179, 181, 183, 185, 187, 189
Social-democratic values, 152
Social organism, 19, 166
Social-practical realities, 186
Social scientific research, 186–187
Social transformation, dynamics, 49
Society for Bioethics and Humanities (ASBH), 123
Society for Bioethics Consultation, 125
Society for Health and Human Values, 125, 169
Socio-biological adaptive strategies, 7
Socio-political agenda, 4, 15
Socio-political environment, 60
"Specialty society", 128
Specification process, 142
Spike, Jeffrey P., 115
Spinoza, B., 25, 159
Spirituality, 55, 62, 136, 178
State-controlled health care, 136
Stem cell research, 124
Stevens, T., 132, 145, 167
Strategically ambiguous appeals, 107–109
considerable consensus, 108
family-based healthcare savings accounts, 109
foundational moral challenge, 108
moral authority of "consensus", 108
moral pluralism, 109
moral truth, 109
traditional religions, 109
unlimited desires, 109
Student unrest in Europe, 9
Sturleson, S., 7
Subramanian, S.V., 66
"Successful bioethicists", 130
Success of bioethics, 18–19, 162–165
Sufficient maturity of child, 103
Susman, Irwin, 37
Swazey, J.P., 14–15, 99–100, 113, 115
Swiderski, D.M., 186–187
Symons, D., 7

T

Tachycardia, 58
Taken-for-granted fashion, 153
Tangwa, G.B., 137
Tao Lai Po-Wah, J., 137
Tarzian, A.J., 115
Teel, K., 169
Teenage sexual activity, 57
ten Have, H., 170
A Theory of Justice, 157
Theory of morality, 5
Thomson, J.J., 57–58
Tollefsen, C., 119
Totalitarian culture, 22
Toulmin, S.E., 34, 36, 41–43
Traditional medical morality, inadequacies, 73
Traditional morality, ideals of, 57
Treaties of Westphalia, 168
Trotter, G., 8–11, 13, 49, 52, 56, 60, 64–65, 85
Tse-tung, Mao, 59

U

Unethical behaviors, 178, 188
Unethical human research, 132
Universal Declaration on Bioethics and Human Rights, 104–105, 107, 111, 117–119
Universal Declaration on Human Rights, 107, 110
Universal secular ethics, 2
U.S. Census Bureau, 61, 140

V

Vatican II, 9
Vattimo, G., 6–7, 25, 158
Veatch, R.M., 73–74, 82, 132
Versenyi, L., 24
Veterans Health Administration (VHA), 183
Vietnam War, 51, 59–60
"Viking morality", 7
Vovelle, M., 168
Vulnerability, 133, 135

W

Waldorf teachers, 123
Walter, J.K., 167
Wang, M., 137
Wang, X., 137
Wascher, W., 61, 67
Well-to-do sick, 77
Wen, Chun-Feng, 137
Whaples, R., 61

Williams, John R., 118
Willkür, 12
Wilson, A.N., 26
Wilson, J., 167
Wilson, Woodrow, 62
Wolfe, T., 64
Wolf, S.M., 46
Wolpe, P.R., 129, 138–139

Woodstock Festival, 9
World Medical Association, 117

Y
"Yellow Book", 44
Yesley, M., 8–9, 31
Youngner, S., 132
Youngsters, counseling, 57